THE LEARNING CYCLE

Elementary School Science and Beyond

REVISED EDITION

Edmund A. Marek
Ann M. L. Cavallo

HEINEMANN
Portsmouth, NH

Heinemann
A division of Reed Elsevier Inc.
361 Hanover Street
Portsmouth, NH 03801–3912

Offices and agents throughout the world

The authors and publisher wish to thank those who have generously given permission to reprint borrowed material.

Excerpts from "Development and Learning" by J. Piaget. From *Journal of Research in Science Teaching* (vol. 2, no. 3, 1964). Reprinted by permission of the publisher, John Wiley & Sons, Inc.

Library of Congress Cataloging-in-Publication Data
Marek, Edmund A.
The learning cycle : elementary school science and beyond / Edmund A. Marek, Ann M. L. Cavallo. --Rev. ed.
p. cm.
Rev. ed. of: The learning cycle and elementary school science teaching / John W. Renner, Edmund A. Marek. c1988.
Includes bibliographical references and index.
ISBN 0-435-07133-5
1. Science--Study and teaching (Elementary)--United States. 2. Learning. 3. Child development--United States. I. Cavallo, Ann M. L. (Ann M. Liberatore) II. Renner, John Wilson, 1924–1991 Learning cycle and elementary school science teaching. III. Title.
LB1585.3.R395 1997
372.3'5044--dc21 96-52276
CIP

Editor: Leigh Peake and Victoria Merecki
Production: Melissa L. Inglis
Cover design: Joni Doherty
Manufacturing: Louise Richardson

Printed in the United States of America on acid-free paper
06 05 VP 7

This book is dedicated to Professor John W. Renner (July 25, 1924–August 10, 1991). Jack's influences in science education are significant and pervasive. More importantly, he touched many lives that were forever changed for the better. Thank you, Jack.

CONTENTS

From the Authors vii

Acknowledgments ix

PART ONE: EXPLORATION

1 The Nature of Science and Science Teaching 3

2 The Goals of Science Education 17

3 The Nature of the Learner 34

PART TWO: TERM INTRODUCTION

4 The Theory Base of Elementary School Science 69

PART THREE: CONCEPT APPLICATION

5 Developing Learning Cycles 105

6 Methods and Technologies Within the Learning Cycle 128

7 Measuring Students' Progress in a Learning Cycle Program 141

8 Learning Cycles for Elementary School Science 152

Appendix A: Learning Cycles for Pre-K and Kindergarten 223

Appendix B: Protocols for Formal Operational Tasks 230

Index 241

FROM THE AUTHORS

Consider the following question: What happens to the white when snow melts? A fourth-grader posed this wonderful question to me many years ago. I immediately realized that a complete, scientifically accurate answer was beyond this nine-year-old's understanding; nonetheless, asking the question was not beyond her imagination. As teachers we must carefully consider *what* children can learn *and how* they learn. This book is designed to guide teachers in developing an understanding of how students learn, as well as to increase their knowledge of the science content that is appropriate for elementary school students.

Robert Karplus, director of the Science Curriculum Improvement Study of the 1960s and professor of physics at the University of California, Berkeley, hypothesized that the teaching of science requires more than content. Teaching requires a plan derived from both the discipline of science and the manner in which students learn. Karplus called the teaching procedure that was invented to satisfy those requirements the *learning cycle*. The learning cycle moves children through a scientific investigation by allowing them first to explore materials, then to construct a concept, and finally to apply this concept to new ideas. This book is about the learning cycle.

The book's format itself models a learning cycle. You, the reader, are invited to explore your own ideas about teaching science, to construct the relationships between the nature of science and the nature of the learner, and finally to expand these ideas into practical applications in your classroom.

You will also discover that the organization of this book goes beyond the discipline of science. For example, the book includes integrated learning cycles and learning cycles across the disciplines. It includes the uses of various questioning strategies, alternative evaluation schemes, and modern technologies. In this book *science* is the *vehicle* for the *processes* of the *learning cycle*. In our formula for teaching in elementary school, the learning cycle is a constant, while the subject—the vehicle—could be language arts, mathematics, social studies, fine arts, or science.

Edmund A. Marek

UPON SEEING A SPIDER, my two-year-old daughter shows great excitement and delight. "Isn't she beautiful, Mommy?" she says.

"Yes, she is very beautiful," I reply.

"What is her name?" she asks me.

"You pick any name you want for her," I tell my daughter.

"Where is her home?" she asks.

"Let's watch her and see," I respond.

Now, there are many responses an adult could give a child upon seeing a spider—especially of the size observed during the above experience! This response was chosen, however, to allow the child to make her own discoveries of nature, and build understandings of how the world works. This response was also chosen to nurture the child's natural curiosity and joy of learning. As children observe nature—and notice how we, as adults, respond to their observations—they construct ideas, understandings, and attitudes about science that will last a lifetime. It is critical that we—as teachers and as parents—ensure that these early learning experiences are engaging and positive.

The learning cycle teaching procedure presented in this textbook is based on this philosophy of science education. The learning cycle, appropriately, places *the children* at the center of learning experiences, with the teacher promoting and encouraging their learning processes. With the teacher's guidance, the children engage in explorations, bring ideas to fruition, form new understandings, and relate them to other concepts in science, subjects across the curriculum, and everyday events. In essence, children's *experiences* allow them to use and develop their thinking processes, and form meaningful understandings of the world.

The formulation of *meaningful understandings* is critically important for children in today's schools. Students need to understand the relevance of their school learning to their own lives and realize its relationship to personal and societal goals. The learning cycle is an instrument that fosters such understanding among school children. Originally founded on Piagetian theory, the learning cycle elucidates several other theories of learning and development, including Ausubel's theory of meaningful learning and Vygotsky's social constructivist theory. The learning cycle is an advanced paradigm that promotes thinking abilities and meaningful learning among children in diverse classroom settings. Moreover, the learning cycle is a model for teaching in *all* subject areas: it provides a basis for thematic and integrated instruction and offers many opportunities to measure authentic learning.

As I continue to observe my daughter look with wonder at the spider weaving her web, I hope that her future will be filled with such enthusiastic discoveries of nature. It is my anticipation that teachers using this book will agree, and promote the interest, curiosity, and excitement toward learning that exist in *all* young schoolchildren.

Ann M. Liberatore Cavallo

ACKNOWLEDGMENTS

Many individuals have contributed to the development of *The Learning Cycle*. We gratefully acknowledge these individuals for their effort and assistance:

The family of Dr. John W. Renner deserves a special thank you for graciously consenting to permit material from the previous edition to be used here. We gratefully recognize the significance of his contributions to this book.

Several teachers provided valuable text related to science teaching and curricula. Special acknowledgments are extended to William T. Fix, Sharlene Kleine, Joseph Green, and John King for their important contributions on implementing various teaching techniques in the learning cycle paradigm. In addition, we gratefully acknowledge David Quine, William Simpson, and Tina McGuffin for their texts on integrated learning cycles and learning cycles in other disciplines.

We also wish to acknowledge several individuals for their assistance with word processing, editing, and proofreading portions of the manuscript. We thank Brenda Peterson, Michelle Borden, Margaret Barbour, Georgianna Saunders, and Stacy Sheperson for their assistance.

Most importantly, we extend our appreciation to our families. Their support during the development of *The Learning Cycle* was invaluable. We lovingly thank you.

PART 1

EXPLORATION

1 The Nature of Science and Science Teaching

Children learn about the *nature of science* from investigating nature. By the nature of science we mean what science is. Science is usually thought of as disciplines—biology, chemistry, geology, meteorology, and physics—that is, content to be taught. Certainly, content from disciplines such as those listed must be taught for the teaching activity to be recognized as *natural* science, that is, the *science of nature.* These disciplines are concerned with explaining nature. But what does the science of nature mean? In other words, what is science?

Albert Einstein stated that "the object of all science is to coordinate our experiences and bring them into a logical system" (Holton and Roller, 1958). How does one learn science by coordinating experiences? First, learners must be allowed to *have* experiences and to coordinate them. In science the laboratory supplies experience—a series of activities that provide data about a natural phenomenon—to the investigator, who then interprets these data, and in so doing, develops an understanding of the phenomenon being investigated. In Einstein's description of science, the laboratory activities provide the "experiences," the interpretation of which leads to the "logical system."

To appreciate fully what Einstein is saying, we will examine the teaching procedure used in teaching science in the *majority* of classrooms throughout the world—the *inform-verify-practice* (IVP) procedure. Usually this procedure begins with a lecture that orally delivers information to students about the science concepts to be learned and the associated terminology. The information and conceptual language can also be delivered through some media such as television, computers, films, or the printed page, but in all these forms, the assumption is made that the language used has meaning for the students. According to this teaching procedure, there need not be any prior experience to make the language meaningful other than a careful definition of terms, which is why textbooks carry carefully prepared vocabulary lists. Regardless of how the information is conveyed, in the first phase of this teaching procedure—*exposition*—the teacher tells the learner what is to be known. Reading has a prominent position in this teaching procedure. In other words, exposition requires that the teacher *inform* the students of what is to be learned.

In the second phase of the traditional teaching procedure, learners usually are shown proof that what they have been told is true. Science includes perhaps the best procedure found in any discipline for students to test the authenticity of what they have been told. Apparatus and materials are available to do the testing. For example, when students are told that a particular chemical reaction is exothermic, they can carry out the reaction and verify it for themselves. Such exercises are sometimes called experiments, but they are not true experiments because the outcome is known before the activity is performed. These activities are simply verification and are extremely important to the exposition teaching procedure. The *verification phase* is the best opportunity students have to attach meaning to the language of the concept they have been given during the information-giving phase.

In the last phase, students may answer questions, solve problems in the textbook, take quizzes, and/or do additional readings. This phase rarely includes any further experience with apparatus and materials, and the activities are usually conducted on the verbal level. This phase gives students an opportunity to *practice* using the information they have been given. Teaching science in most schools is done with the inform-verify-practice (IVP) procedure.

Focusing attention on Einstein's description of science immediately reveals the first deficiency of the IVP teaching procedure. Students have no experiences to coordinate; they are *informed* about what they are to know. The experiences *someone else* has had are coordinated into a logical system and presented to them. The IVP teaching procedure tells students that science is a finished procedure—here are its products—that they are expected to know. Do not believe that the experiences students have during a verification laboratory are the kinds of experiences Einstein was describing. In a verification laboratory, students simply reenact with materials—apparatus, chemicals, living things—what the textbook tells them. The verification laboratory is further disqualified as a science experience because students know the outcome all of the time the "laboratory" is in session. If Einstein is correct, we must claim that science cannot be taught with the IVP teaching procedure.

Maria Mitchell, a nineteenth-century astronomer, succinctly captured the way students do learn science when she said, "Students should question what they are taught and learn for themselves" (Mack, 1990). Physicist Niels Bohr, who left his imprint on science by inventing the Bohr model of the atom, stated that "the task of science is both to extend the range of our experience and reduce it to order" (Booth, 1962), a description closely parallel to Einstein's. All these scientists are concerned about experiences with phenomena and organizing those experiences so they make sense to us.

Critics of the idea that science is a process of inventing explanations for phenomena found in the natural world will always say, "Yes, but where do the facts of science fit into that description of science?" The nineteenth-century French scientist Henri Poincairé addressed that question when he said, "Science is built up with facts, as a house is with stones, but a collection of facts is no more a science than a heap of stones is a house" (Kelly, 1941). Poincairé left little room for the argument that memorizing and being able to repeat the facts of science represents being educated in science; these activities—memorizing and repeating—do not represent science. Furthermore, teachers who concentrate on leading students to memorize the facts, laws, and principles of science—the IVP procedure—are not teaching science; the students have only the stones, not the house. To be educated in science, students must have experiences that they reduce to a logical system. They have then learned how to construct knowledge about the natural world—they have learned science. According to Poincairé, they have built the house.

This emphasis on the *processes* of science is not unique to this discipline. Winifred Edgerton Merrill, the first woman to earn a degree in mathematics at Columbia University, described her work as "this mental search for coordinating elements in life experiences, in art forms, in the complexities of educational problems, always searching for a better understanding of the nature of things through some underlying unifying principle" (Green and Laduke, 1990).

RECALL YOUR OWN EXPERIENCES IN THE SCIENCE CLASSROOM AS A STUDENT. WHAT FORMAT—TEACHING PROCEDURE—WAS USED? WHAT ASPECTS OF THE FORMAT STRENGTHENED YOUR UNDERSTANDING OF SCIENCE AND LEARNING? WHAT ASPECTS OF THE FORMAT DETRACTED FROM YOUR UNDERSTANDING OF SCIENCE AND LEARNING?

Science, then, is a process of finding facts, laws, principles, and concepts, but the content does not represent science. Perhaps science historian Duane Roller said it best: "science is the quest for knowledge, not the knowledge itself" (Roller, 1970). Teaching what is called science without involving the students in a quest or search is not teaching science. Let's examine the quest that Roller mentions in his description of science. We will do this by conducting an investigation and then using the investigation as a model to illustrate the processes of science.

EXPLORATION: You will need a wire, a dry cell, and a lightbulb. The objects are shown in the picture. Make a system of the three objects (see Figure 1–1). What evidence do you find that the objects will interact? After you find out what happens, draw a picture of how you connected the wire, the dry cell, and the lightbulb.

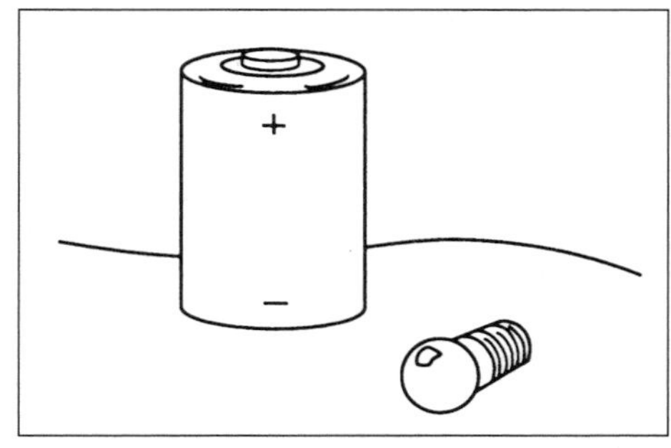

FIG. 1–1

First, use one wire to produce interaction. Then use two wires. Draw a picture of how the two wires, the dry cell, and the lightbulb were connected when you saw evidence of interaction.

Compare the two pictures you have drawn. How are they alike? How are they different?

At how many *different* points did you have to touch the dry cell before you observed evidence of interaction? At how many places did you touch the lightbulb before you saw evidence of interaction?

Each of the six pictures in Figure 1–2 shows a system made up of a wire, a dry cell, and a lightbulb. The objects in the system are connected differently in each picture. Why does the system remain the same system?

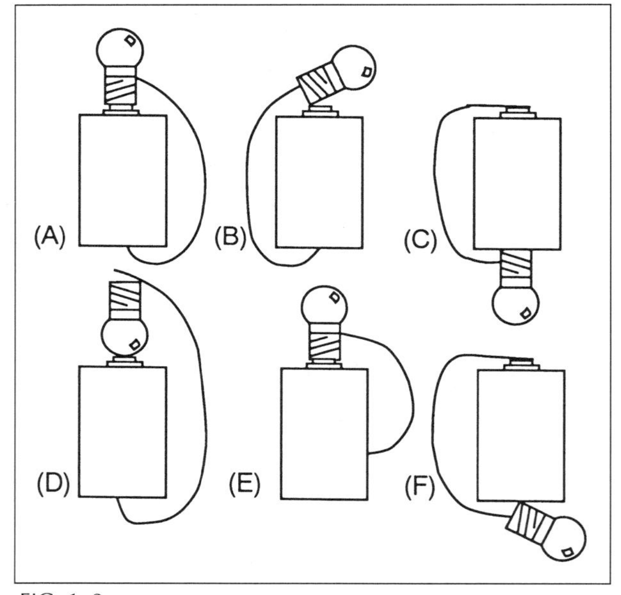

FIG. 1–2

Study the various ways in which the objects in the system are arranged. Predict which of the arrangements will produce evidence of interaction. Record your prediction in a table like the one shown in Figure 1–3. Each end of a dry cell is called a pole. Note that one pole is marked "+" and the other is marked "–." The "+" stands for *positive* and the "–" stands for *negative*.

You are predicting whether each of the arrangements of the objects in the system will give evidence of interaction. Write "yes" or "no" in your table to show whether or not you expect to see evidence of interaction.

Earlier, you answered questions about the number of places a dry cell and a lightbulb had to be touched before you observed evidence of interaction. Refer back to the answers to those questions. If you wish to change any of your predictions, do it now.

From now on, do not change any of your predictions. You are going to do experiments with a wire, a dry cell, and a lightbulb to test your predictions. Arrange the objects in the system just as they are shown in pictures A, B, C, D,

Picture	Prediction (yes or no)	Experiment (yes or no)
(A)		
(B)		
(C)		
(D)		
(E)		
(F)		

FIG. 1–3

E, and F. Record in your table whether or not you found evidence of interaction by writing "yes" or "no."

Do not be concerned if you predicted incorrectly for some arrangements of the objects. Figure out why you predicted incorrectly. Record what you decided about incorrect predictions.

The foregoing directions for the exploration phase of this teaching procedure are quite specific in directing children how to interact with materials. Take careful note that nowhere in those directions are the words electrical current and electrical circuit used, nor are the students told what concepts they are to find. The directions lead the students to gather data that let them absorb the essence of the concepts. The directions—and particularly the data in the table shown in Figure 1–3—do one additional thing: they lead students to find several contradictions between the "Prediction" and "Experiment" columns. Those contradictions lead them to ask, "Why did that happen?" "Will it happen again?" and so on. Such questions are evidence that the students are now ready to be led to construct the concept to be learned.

The teacher next asks the students to contribute the data they have gathered to a class discussion. The teacher and the students put all these data together and construct an explanation that represents the concept to be learned. The teacher then supplies the appropriate terminology or vocabulary if the students do not know the terms. This phase of the teaching procedure is called *term introduction* (Lawson, 1988). An example of what might happen during a *teacher-led discussion* in the term introduction phase follows.

TERM INTRODUCTION: A wire, a dry cell, and a lightbulb can be arranged in a system that causes an interaction—the lightbulb lights.

With the lighting of the lightbulb, energy is being used. The dry cell supplies the electrical energy. But in order to see evidence of interaction, you had to arrange the wire, the dry cell, and the lightbulb in a certain way.

The wire, the dry cell, and the lightbulb were arranged in a special way. With this arrangement you observed evidence of interaction. Such an arrangement is called an *electrical circuit.*

The lightbulb will not light without the dry cell. The dry cell is the energy source in the circuit. The *interaction* of the wire, the dry cell, and the lightbulb causes something to happen in the circuit. That interaction causes

electrical current.

Any evidence of electrical interaction requires current. When the starter on a car begins to hum, current is there. When a refrigerator, a computer, or a television operates, current is also present. Current is present when an electric light is turned on.

People sometimes refer to electric current as "juice." The word "juice" suggests a mental picture. Those who use the word are trying to make a model of electric current.

The students have acquired two new concepts. In some investigations one concept is acquired at a time, but in this case, two concepts are inherent in the same materials. Usually there is no surprise when the concept is named. The students expect some unifying idea and have probably been referring to the circuit as "thing" or some other suitable word. The students absorbed the concept before they identified it. When the identification was made known, they were probably not surprised.

The students are now ready to integrate these new concepts with other ideas. They expand the new ideas they have just acquired to include other ideas. We have called this phase of this teaching procedure *concept application*. We have found that using student directions for this phase is helpful and gets the concept application underway. Once this phase has begun, students often suggest other ideas that contribute greatly to their retention of the newly acquired concept. It is extremely important that, during the concept application, the scientific terminology of the new concept be used.

CONCEPT APPLICATION: You will need a dry cell, a lightbulb, two brass clips, three pieces of wire, a holder for the dry cell, and a socket for the lightbulb.

Put the objects together as shown at the left of Figure 1–4. As you will observe, the lightbulb gives no evidence of interaction. Next, touch wire A and wire B together. What happens?

Put different kinds of objects between wire A and wire B. Touch the two wires to the objects (Figure 1–4). Which objects permit electrical current to light the bulb?

When the current lights the bulb, the circuit is a *complete* circuit. Make a record of the kinds of materials you used in your circuit. Indicate which materials made a complete circuit.

Refer once again to the pictures in Figure 1–4. When a system of electrical objects is arranged like the system in the first illustration, the arrangement is called an *open circuit*. The circuit is a *closed circuit* when something is put between wire A and wire B and the lightbulb lights.

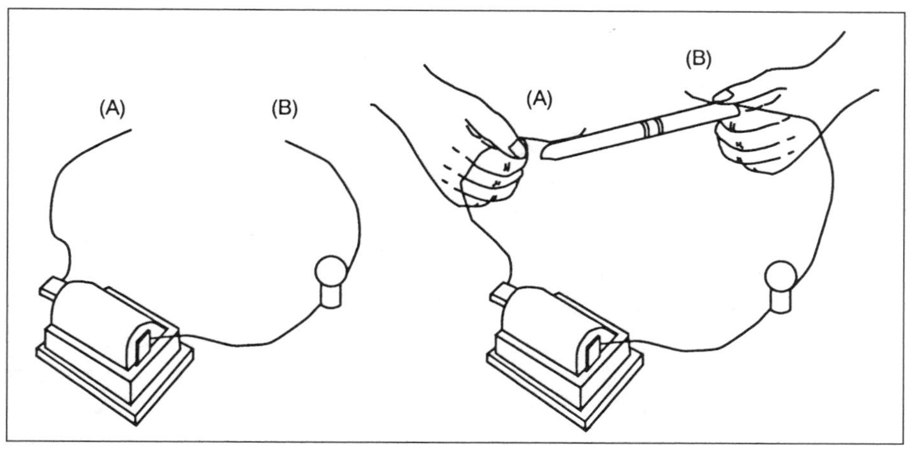

FIG. 1–4

These three phases—exploration, term introduction, and concept application—represent the learning cycle. In the first phase, exploration, the students are permitted to gather their own data and fully explore the materials; this phase is student-centered. The teacher provides directions and materials, answers and asks questions, gives hints and clues, helps with repairing materials and equipment, and generally keeps the exploration going. The term introduction phase is teacher-centered. The teacher establishes the discussion environment, asks for and accepts the data gathered, suggests an organization plan for the data if necessary, and ultimately introduces the scientific terminology for the concept. The roles of the teacher and students in the concept-application phase are like those in the exploration phase; in addition, the teacher must use the terminology of the concept and *insist* that the students use it also.

The experiences students must have in doing science, according to Einstein and Bohr, must provide information—a precise description of the exploration phase of the learning cycle. The information gained from the exploration phase must be formed into a "logical system," according to Einstein, or reduced to order, according to Bohr. Both are saying that in science, the information gained from exploration must be conceptually organized by the investigator. This role of the investigator is echoed by Merrill in her "mental search . . . for a better understanding of the nature of things." In other words, the practice of science requires that the investigator develop conceptual understandings. The phases of the learning cycle, therefore, have their roots in the discipline itself. These phases and their sequences, according to Einstein, Bohr, and Merrill, are science.

SELECT A SCIENCE CONCEPT SUITABLE FOR A GRADE YOU WANT TO TEACH. PREPARE A LEARNING CYCLE TO TEACH THAT CONCEPT. WRITE OUT ALL DIRECTIONS.

Concepts, Facts, Generalizations, and Principles

The atomic weight of sodium is 22.9898 atomic mass units. Wood comes from trees. Water will not dissolve oil. Oak trees have green leaves in the summer. In the winter the temperature of the water in a lake is warmer at the bottom than at the top. All of those statements are *facts*. A fact is something that actually exists. Because of their actual existence facts can be checked. You weigh a definite amount, you are a specific age and height—these are facts because they can be checked by someone else and similar results found. Facts do not *by themselves* lead to the construction of knowledge; they are end points. Do not interpret this last statement as being derogatory; facts are very much needed in the learning process. When we are learning something new, we probably assimilate facts or what we think are facts. That assimilation, however, does not represent learning. We believe, however, that the assimilation of facts starts the entire learning process, but learning has not taken place if all students are asked to do is to "know" the facts.

A child places a bar magnet on a flat, paper-covered surface and slowly pushes small metal pieces toward it. The metal pieces reach a certain point where they no longer need to be pushed; they jump toward the magnet. The young investigator makes a pencil mark on the paper at the point where the metal piece jumped. After a time, there are many such marks surrounding the magnet, outlining an oval-shaped area. The child-investigator is asked to explain what the data from the investigation mean. The young experimenter states that, if a small piece of the metal being used is placed in the space outlined around a magnet, the metal piece jumps toward the magnet. That statement is the concept. The child has obviously conducted an exploration, and the data from that exploration have led the child to develop a summarizing statement. *The child has just constructed a concept.* To complete the concept development, the teacher needs to explain that the space the child outlined with the pencil marks is called the *magnetic field*. The teacher provides the terminology. Please remember, however, that stating the name of a concept is *not* stating the concept. In listing concepts to be taught, stating the names of the concepts is not sufficient. For a concept-application experience, the child could use other bar magnets or magnets of other shapes to outline magnetic fields.

The learning cycle just described tells how a concept is formed but does not explain what a concept is. Consider what the child in the magnetic-field investigation did. First, the child collected facts: (1) the metal piece jumps toward the magnet; (2) the metal piece jumps toward the magnet at a particular point; (3) the metal piece outside that point has to be pushed toward the magnet. From these experiences, the child abstracts the idea that there is a specific space around the magnet in which the metal pieces jump toward the magnet. In other words, a concept is a specific idea abstracted from particular instances. How does that concept differ from a *generalization?* It probably doesn't, but spending the time necessary to pinpoint the specific similarities and differences, if any, hardly seems worthwhile. A concept is abstracted from experience, and a generalization can be a proposition that need not come from experience. This is the only meaningful difference we see, but even that is not too meaningful. Concepts are put together—abstracted—from data that came from experience by the *process* of generalization. Our belief is that *knowledge is constructed by abstracting concepts from some type of experience.*

Throughout the foregoing discussion, we have referred to the process of "abstracting" concepts. Do not confuse this use of abstract with abstract subject matter. Think of "to abstract" or "abstracting" as a process, and abstract subject matter as a body of content, such as higher mathematics or music theory.

Teachers frequently speak of "teaching concepts." This is not possible, since concepts are learned. The responsibility of the teacher is to provide the materials, the environment, and the guidance needed to assist students in constructing their own concepts about particular phenomena in science. In other words, concepts are what students carry away from a learning experience. A perfectly legitimate label for learning content is *conceptualization.*

The concepts on which the school's curriculum is based, therefore, are those we wish students to learn. These concepts represent the *only* content learned. We as teachers must state the concepts before constructing the curriculum. We must keep in mind, however, that they are our concepts, and we are assuming that our concepts are harmonious with the content being taught. We also hope that at the end of a learning experience, the concepts of the students match ours and the content. *Concepts,* therefore, *are the ideas about a particular phenomenon students abstract from a learning experience.*

Consider this statement: *Every electric current has a magnetic field surrounding it.* Now this statement has meaning only if the person encountering it understands the concepts of electric current and magnetic field. The statement relating those two concepts is a *principle,* which the dictionary defines as the laws or facts of nature underlying the working of an artificial device. To understand how an "artificial device" works, one must understand the concepts on which it is based. It is such a statement of conceptual relationships that constitutes a principle.

WRITE ONE SCIENCE CONCEPT FOR EACH GRADE IN ELEMENTARY SCHOOL— KINDERGARTEN THROUGH GRADE SIX—AND DEFEND WHY EACH STATEMENT YOU WRITE IS A CONCEPT.

Consider the concept of temperature. Second-grade children—seven-year-olds—can develop the understanding that when something hot is brought in contact with a thermometer, the liquid in the thermometer rises. In other words, the "hotness" of an object is the reading on a thermometer. By fourth grade (nine-year-olds), children understand the concept that the *temperature* of an object is represented by the reading on a thermometer. This is a perfectly workable concept of temperature and in some taxonomies of concepts is labeled a *concrete* concept. Temperature, however, can also be thought of in terms of the kinetic-molecular theory of matter, which states that all molecules in an object are in motion as long as the temperature remains above absolute zero. The theory also states that as the temperature of an object rises, the object's molecules increase their motion. If temperature is considered in terms of the kinetic-molecular theory, it is labeled an *abstract* concept because understanding it requires reasoning according to this postulate.

Therefore, concepts can be identified as concrete or abstract depending on how they are developed by learners. The criterion for judging whether or not a concept is concrete is very direct. If a concept can be constructed from data collected through *direct experience* with objects, events, or situations, the concept is concrete. If an assumption, postulate, or axiom is used to collect or interpret data leading to the conceptual understanding, the concept is abstract.

Some abstract concepts cannot be made understandable at a concrete level. Consider the atom as an example. To understand the atom and its structure a student must understand many concepts—such as electrical charge—that cannot be directly experienced. Even though much *indirect* evidence exists to support these concepts, they are postulates.

Teachers frequently use models to help students understand difficult concepts. One model, and one of the most widely used, is the atomic model. Small spheres and sticks are frequently put together to represent not only the entire atom but its nucleus and the electrons in orbit around the nucleus. Do such models assist students in learning abstract concepts? Those models are extremely helpful in assisting *abstract-thinking* students to learn abstract concepts because a model is really a postulate—in this case, a belief about what something looks like. The concrete thinker cannot reason with postulates and therefore does not understand that the model is a representation of what we believe the atom looks like. Since concrete learners can deal only with reality, they learn the model but do not understand that it is only a model. They will not understand any questions asked of them or problems given to them about what the model represents because these do not deal directly with the model. Models do not necessarily provide *direct* experience with the phenomena, so concrete learners do not learn abstract concepts from them.

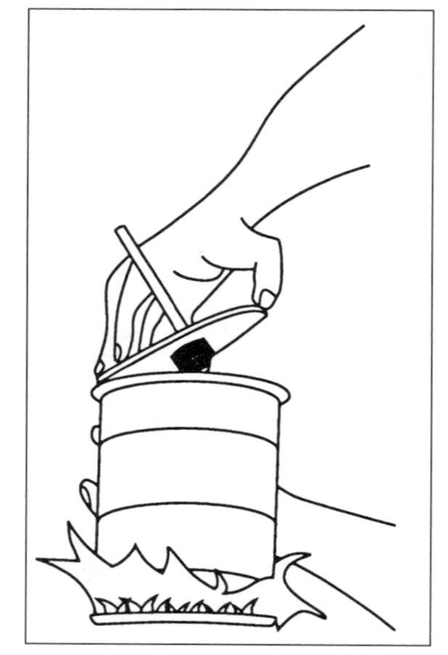

FIG. 1–5 *A breathing model*

Now here is an exception to the above statement about models. You have probably made a model (Figure 1–5) showing how we breathe using a coffee can, a one-holed rubber stopper, a glass tube, a small piece of rubber, and a balloon. A one-liter, plastic soda bottle substitutes well for the coffee can. The bottom is cut from the can, or soda bottle, and the rubber piece is stretched over the open bottom and held securely around the can's sides by a rubber band. A hole is made in the can's plastic top, and the rubber stopper, through which the glass tube has been passed, is sealed in the hole. A balloon is attached to the end of the glass tube, which will be inside the can. The rubber piece is pinched at its middle and pulled outward.

The pressure inside the can drops due to an increase of volume inside the can, and outside air pressure forces air into the balloon through the end of the glass tube exposed to the outside air. This model accurately depicts air movement due to pressure differences similar to our breathing process. When we expand our body cavity, the pressure inside drops. The outside air pressure forces air into our lungs through our nose or mouth. Now if children can watch the model work and at the same time place their hands on their ribs and feel their body expand, they will be provided with *direct experience*; the model helps only to explain that experience. When a model augments direct experience, it is useful to concrete thinkers.

The Results of Concept Misplacement in the School

A concept that is frequently found in elementary school science is dinosaurs and why those creatures are no longer found on the earth. The disappearance of the dinosaur is an abstract concept because it involves stating hypotheses that cannot be tested. Obviously, direct or concrete experience with the concept is impossible, and models are not helpful. To find out what senior high school students do understand about dinosaurs, we designed and carried out a study. Our results showed that only 4 percent of the students participating demonstrated sound understanding of the concept related to the disappearance of dinosaurs. A common misunderstanding was that "human beings killed the dinosaurs either directly or indirectly, for example by destroying their food or habitat or by polluting their water." This study, we believe, generates a hypothesis about the results of requiring concrete students to study abstract concepts, which can be stated as follows: When students are required to attempt to learn concepts above their intellectual levels, they develop misunderstandings. We will discuss more of our research into misconceptions in Chapter 7.

NOW IT IS YOUR TURN. WRITE DOWN TWO CONCEPTS THAT ARE CONCRETE AND TWO THAT ARE ABSTRACT. EXPLAIN WHY THE CONCEPTS ARE CONCRETE OR ABSTRACT. NOW WRITE DOWN TWO CONCEPTS THAT CAN BE LABELED AS CONCRETE *OR* ABSTRACT CONCEPTS. EXPLAIN HOW THE CONCEPTS CAN BE EITHER CONCRETE OR ABSTRACT.

The Learning Cycle as a Foundation for Elementary School Science Programs

The learning cycle teaching procedure was originated by Professor Robert Karplus, a physicist at the University of California at Berkeley, in the late 1950s and early 1960s. Karplus had been a guest teacher in his daughter's second-grade class, and presented concepts on electrical charges. This visit led him to discover that children had difficulty learning science concepts. He continued to teach science to second- and third-graders for several years. These classroom experiences prompted Karplus to turn his thoughts toward developing a program for elementary school science (Lawson, Abraham, & Renner, 1989). The program that resulted from the efforts of Karplus and a team of scientists and educators was designed to be consistent with the discipline of science—that is, to match the investigative steps that scientists have used throughout history in the formulation of new inventions and theories. In 1967, Karplus and Thier identified and titled three phases important for teaching science: "preliminary exploration," "invention," and "discovery." These phases have since been renamed, as described in this chapter, "exploration," "term introduction," and "concept application," respectively. The teaching procedure developed by Karplus and Thier (1967) was the learning cycle.

The learning cycle served as the foundation for the Science Curriculum Improvement Study (SCIS) program, which was first developed in the 1960s through a project supported by the National Science Foundation. In the 1970s, several trial teaching centers were identified for classroom testing of the SCIS program. One trial center was the University of Oklahoma, with the Norman Public School District as the test site, under the direction of Dr. John W. (Jack) Renner. The Norman public schools and many other school districts throughout the country continue successfully to use the learning cycle and new versions of the SCIS program (including SCIIS and SCIS-3) as their elementary school science program to this day. Why are the learning cycle and curricula derived from it so successful? Research that has evaluated the SCIS program has found that students who have used it are superior to non-SCIS students in both cognitive and motivational behavior (Kyle and Bonnstetter, 1992). Thus, the students of teachers who use the learning cycle have greater success and achievement in science and are more motivated to learn science than students whose teachers use other instructional procedures.

Given its great impact on student learning and achievement, the learning cycle has served as the origin and foundation for several other science curricula currently in use. Two of these curricula are Biological Science Curriculum Study (BSCS) and Full Option Science System (FOSS). As an example, the FOSS curricula developers state the following in the introductory section of their curricula:

> A learning cycle is embedded in FOSS activities. Activities often start with free exploration of materials, followed by a discussion of discoveries in which vocab-

ulary is developed and ideas clarified . . . followed by additional experiences with materials. (Regents, University of California, 1993)

The basis for the FOSS curricula seems to be consistent with the learning cycle teaching procedure as it is described in this text.

It is important to be aware that developers and users of the learning cycle hold a teaching philosophy known as *constructivism*. This teaching philosophy contends that learners must actively formulate or *construct* understandings for themselves, based on their experiences. (We will discuss learning models in depth in Chapter 3). This philosophy is in contrast with more traditional views of teaching in which students are seen as passive receivers of information. It is critical that current and future teachers evaluate with much scrutiny the curricula they use and the teaching philosophy that serves as their foundation. The curricula must *match* their purported constructivist philosophy. The curricula cited in this section use the learning cycle, and therefore do match and support the constructivist philosophy of teaching—but teachers need to evaluate the extent of the match and the overall quality of each curriculum for themselves.

REFERENCES FOR CURRICULA BASED ON THE LEARNING CYCLE ARE PROVIDED AT THE END OF THIS CHAPTER. REVIEW THESE CURRICULA, ALONG WITH OTHERS NOT BASED ON THE LEARNING CYCLE TEACHING PROCEDURE. COMPARE AND CONTRAST THESE CURRICULA IN TERMS OF THEIR CONSISTENCY WITH THE NATURE OF SCIENCE. HOW CLOSELY DOES EACH CURRICULUM FOLLOW THE LEARNING CYCLE? HOW HAS THE LEARNING CYCLE BEEN MODIFIED AMONG DIFFERENT CURRICULA? WHAT TYPE OF LEARNING OUTCOMES WOULD YOU EXPECT AMONG STUDENTS FROM LEARNING CYCLE CURRICULA AS COMPARED TO NON–LEARNING CYCLE CURRICULA?

References

BOOTH, V. H. 1962. *Physical Science*. New York: Macmillan.

GREEN, J., AND J. LADUKE. 1990. "Contributors to American Mathematics: An Overview and Selection." In *Women of Science*, edited by G. Kass-Simon and P. Farnes, 117–146. Bloomington and Indianapolis: Indiana University Press.

HOLTON, G., AND H. D. ROLLER. 1958. *Foundations of Modern Physical Science*. Reading, MA: Addison-Wesley.

INHELDER, B., AND J. PIAGET. 1958. *The Growth of Logical Thinking*. New York: Basic Books.

KARPLUS, R., AND H. D. THIER. 1967. *A New Look at Elementary School Science*. Chicago: Rand McNally.

KELLY, H. C. 1941. *A Textbook of Electricity and Magnetism*. New York: John Wiley and Sons.

KYLE, W. C., AND R. J. BONNSTETTER. 1992. *Science Curriculum Improvement Study: Profile of Excellence.* Hudson, NH: Delta Education.

LAWSON, A. E. 1988. "A Better Way to Teach Biology." *American Biology Teacher* 50 (5): 266–289.

LAWSON, A. E., M. R. ABRAHAM, AND J. W. RENNER. 1989. *A Theory of Instruction: Using the Learning Cycle to Teach Science Concepts and Thinking Skills.* NARST Monograph, Number One.

MACK, P. 1990. "Straying from Their Orbits: Women in Astronomy in America." In *Women of Science*, edited by G. Kass-Simon and P. Farnes, 117–146. Bloomington and Indianapolis: Indiana University Press.

REGENTS, UNIVERSITY OF CALIFORNIA, 1993. *Full Option Science System.* Chicago: Encyclopedia Britannica Educational Corporation.

ROLLER, D. 1970. "Has Science a Climate?" *Sunday Oklahoman*, February 22, 1970.

Curriculum Resources

BIOLOGICAL SCIENCE CURRICULUM STUDY (BSCS). 1992. Kendall Hunt Publishing Company, 2460 Kerper Boulevard, P.O. Box 539, Dubuque, IA, 52004.

FULL OPTION SCIENCE SYSTEM (FOSS). 1993. Encyclopedia Britannica Educational Corporation, 310 South Michigan Avenue, Chicago, IL 60605. 1–800-554-9862.

SCIENCE CURRICULUM IMPROVEMENT STUDY (SCIS, SCIIS, SCIS-3). 1970–1992. Delta Education, P.O. Box 915, Hudson, NH 03051. 1-800-258-1302.

2 *The Goals of Science Education*

> Natural science, in its relations to practical life, may almost be said to have been born in this century; yet so rapid has been her growth that she has already transformed society and revolutionized the habits of thought. Science has fearlessly knocked at every door; and where she has been admitted, labor has been saved, time lengthened, distance made less, comfort and health and happiness increased, and the human spirit freed from many a tyranny. (Alling, 1881)

This quotation may appear to have been written recently, but in fact was written over one hundred years ago by a very perceptive educator, Mary Alling. She and others during this time were beginning to realize the impact science was having on individuals' ways of thinking, or "habits of thought." Compare the above passage with one written in the more recently published book *Science for All Americans*, which outlines goals of Project 2061—goals for science education for the twenty-first century.

> Scientific habits of mind can help people in every walk of life to deal sensibly with problems that often involve evidence, quantitative considerations, logical arguments, and uncertainty; without the ability to think critically and independently, citizens are easy prey to dogmatists, flimflam artists, and purveyors of simple solutions to complex problems. . . . The life-enhancing potential of science and technology cannot be realized unless the public in general comes to understand science, mathematics, and technology and to acquire scientific habits of the mind. (American Association for the Advancement of Science, 1990)

There are striking similarities between the two quotations. Throughout the last two centuries science has become an increasingly prominent factor in improving the quality of life and our understanding of the world. Thus, educators, scientists, and others have recognized the impact of science on developing the "habits of the mind," or our ways of thinking. But what are the habits of the mind referred to in these quotations? Are these habits specific to science or are they more general and applicable in a variety of contexts? How should teachers help their students develop such habits of the mind and why? These questions will be addressed in this chapter.

The Purpose of Schools

In order to better understand the goals of science education, as the chapter title indicates, one must first analyze the purpose of education and schools in a more general sense. Many documents have been developed over the past century that discuss the fundamental purpose of schools. None, however, is as succinct in defining that purpose as *The Central Purpose of American Education* by the Educational Policies Commission (EPC), published in 1961. According to the EPC,

> The purpose which runs through and strengthens all other educational purposes—the common thread of education—is the development of the ability to think. This is the central purpose to which the school must be oriented if it is to accomplish either its traditional tasks or those newly accentuated by recent changes in the world. . . . Many agencies contribute to achieving educational objectives, but this particular objective will not be generally attained unless the school focuses on it.

The message of this passage is quite relevant for the education of children in today's schools. Then and now, the central purpose of every school activity should be to lead students to develop their thinking abilities. According to the EPC, "the school is an intellectual institution, and developing the intellect of the students is its central business." Project 2061 also emphasizes the development of thinking as the central purpose of education: "Education . . . should help students to develop the understandings and habits of mind they need to become compassionate human beings able to think for themselves and to face life head on" (AAAS, 1990). The ability to think logically and use reason in life situations are the "habits of the mind" referred to in these documents. Educators need to help children—who represent the future leaders and decision makers of our society—develop the ability to think logically.

Why is it important to have the development of thinking ability as the *central* purpose of education? First, all other goals of education, such as understanding the subject matter, cannot be achieved without the ability to think logically. Second, this goal prepares students for their future in society. Throughout a lifetime, individuals must solve many problems, make judgments and decisions, and, ideally, create new ideas that extend and/or improve current knowledge of the world. These activities will not be accomplished if the students in our schools have not developed the ability to think. Third, developing the ability to think is important across all subjects in the curriculum. Content is specific to subject areas and may be forgotten. The ability to think logically is general and central to all subject areas and will prevail for a lifetime. Finally, individuals in our society cannot be truly free if the ability to think is lacking. How is thinking related to freedom? The ability to think allows individuals to decide, for themselves, the value of others' decisions, opinions, and rules. Without the ability to think, reason, and form opinions independently, individuals would have no choice but to accept the decisions of virtually anyone who is in a position of authority. Note that the two quotations presented at the beginning of this chapter, written over a century apart, emphasize that thinking ability is critical for preserving our freedom.

This final point regarding freedom is perhaps of greatest significance. This nation was founded on the premise that all persons are entitled to their individual freedoms. Freedom, however, requires certain factors for its establishment and survival, and these include "the social institutions which protect freedom and the personal commitment which gives it force" (EPC, 1961). But social institutions will neither be free nor advocate freedom if those governing them do not so demand, and these individuals will not demand freedom if they are not committed to it. In order to demand and practice responsible freedom, individuals must have what the EPC called "freedom of the mind," "a condition which each individual must develop for himself." To be truly free, and to maintain the democratic society we cherish, individuals must use thinking skills that allow each to formulate well-founded opinions, judgments, and actions. Thus, "a free society has the obligation to create circumstances in which all individuals may have the opportunity and encouragement to attain freedom of the mind" (EPC, 1961).

These *circumstances* can be created in our school classrooms. Schools are supported by our free society to perpetuate the principles upon which the society is founded. In order to perpetuate a *free* society, however, the individuals making it up must have freedom of mind. To have freedom of the mind students must learn to think autonomously. If schools are to achieve their *central* purpose, the experiences they provide must lead students to develop the ability to think.

What is thinking? What does having the ability to think enable one to do? The Educational Policies Commission defined "thinking ability" in terms of certain rational powers of the mind. These rational powers are *recalling*, *comparing*, *inferring*, *generalizing*, *deducing*, *classifying*, *analyzing*, *imagining*, *synthesizing*, and *evaluating*. The rational powers enable individuals to apply logic and the available evidence to their ideas, attitudes, and actions and to pursue their particular goals. These rational powers, according to the EPC, are not "all of life or all of the mind, but they are the essence of the ability to think" (1961).

How can teachers foster the development of thinking abilities in students? The rational powers are believed to be natural, innate abilities that must be developed or enhanced through use. Used in concert, these powers define what is called logical thinking. It is possible to separately enhance certain rational powers, such as recall, but separate use of these powers does not constitute logical thinking. Logical thinking is a process that must be learned and practiced. In order to develop one's logical thinking abilities, the rational powers must be used in a coordinated, systematic way. Therefore, teachers must create a classroom environment infused with experiences that allow students to use their rational powers in a coordinated way. Such experiences will consequently develop their students' logical thinking abilities.

The experiences teachers should provide to promote students' development of logical thinking have a simple basis: inquiry. Inquiry is the process that utilizes the rational powers and process skills to learn about some aspect of the world. Science is, by its nature, inquiry. Therefore, science is a natural vehicle to accomplish the central purpose of education.

Science and the Achievement of the Central Purpose

Many individuals view science as a compilation of facts. Teaching too often reflects this belief, evident in the overuse of science textbooks and lectures. In fact, research in science education has reported that the vast majority of teachers use the textbook as their primary teaching tool virtually all of the time (National Science Teachers Association, 1994). Many of these teachers use the science textbook simply to "cover content," or subject matter. Teachers often emphasize rote memorization of such content and do not challenge students to formulate deep understandings of science concepts and information. Importantly, using the textbook as the sole source of science knowledge, and/or telling children facts about the world to memorize will not help them develop thinking skills. The overuse of these expository teaching procedures in science *contradicts the very nature* of science as inquiry, as discussed in Chapter 1. In a later report of the Educational Policies Commission (1966), the authors wrote, "The view of teaching as the indoctrination of superior knowledge and wisdom here gives way to a concept of teaching as promotion of the development of the learner from within." This statement means that children cannot simply be given information and expected to replicate the information as would a parrot. Teaching for rote learning is a grave underestimation of the capabilities of the human mind. Children must use their thinking abilities (rational powers) to construct understandings about the world for themselves. In the process of constructing understandings, children *develop* their thinking skills and acquire knowledge that is meaningful to them. The learning, and the understandings children subsequently acquire, are much greater and *deeper* than can occur through rote learning. If science is taught as inquiry and as a process, students will use less rote learning and more thinking processes in the construction of meaningful understandings of science concepts.

So, how should teaching of science *content* be accomplished? Content is important and is the basis for school curricula. Teachers must possess a thorough knowledge of content because it is what they use to lead students to develop their rational powers. But just the study of subject matter does not "in and of itself . . . necessarily enhance the rational powers" (EPC, 1961). That students know the facts, concepts, principles, and laws of science does not necessarily ensure that they have developed their rational powers. Students who are, for example, "perceiving and recognizing patterns in a mass of . . . data" are learning to analyze, deduce, and infer. That, of course, is science. But the EPC also said the following:

> No particular body of knowledge will of itself develop the ability to think clearly. The development of this ability depends instead on methods that encourage the transfer of learning from one context to another and the reorganization of things learned. (1961)

It is the *way* the content is used and not necessarily the content itself that leads to rational power development. According to the EPC, the teaching procedure we use in science must facilitate both the transfer of knowledge from context to context and the reorganization of knowledge if that teaching procedure is to lead students to develop their rational powers. Thus, students can learn facts, concepts, principles, and laws of science by *directly engaging* in science processes that require the use of their rational powers. In doing so, the students will not only become proficient in the use of these rational powers, but will remember and make sense of the associated concept. The way in which students directly engage in both the process and the content of science is through inquiry.

Using Inquiry to Develop the Rational Powers of the Mind

Inquiry

Inquiry may be described as a search for information, a quest for knowledge, or an exploration of certain phenomena to understand the world better. Scientists and educators have written for many years about the need to involve children in our schools in scientific inquiry. In a speech given in 1869, scientist Thomas Huxley stated that elementary school science teaching must be done through "object teaching" or direct experiences using inquiry with objects. Huxley advised as follows:

> Don't be satisfied with telling him that a magnet attracts iron. Let him see that it does; let him feel the pull of the one upon the other for himself. And, especially, tell him that it is his duty to doubt until he is compelled, by the absolute authority of Nature, to believe that which is written in books.

In other words, children should have experiences with objects, phenomena, and/or nature that raise questions that begin a process of inquiry. The children must use their *minds* to explain their observations and experiences for themselves and not simply believe or "blindly" accept what is written in textbooks. Those who have taught science this way, according to Huxley, have "created an intellectual habit of priceless value in practical life."

Others throughout history have also urged the use of inquiry in teaching science. In 1903, Katherine Camp of the University of Chicago Laboratory School urged teachers to use inquiry and inductive reasoning in their science instruction in the elementary schools. In her 1903 publication, Camp posed the following questions in discussing the form elementary science instruction should take to best help children learn science:

> Might not the ideal [learning] be stated as power to be gained through observation, accompanied, or followed at a respectful distance, by inductive reasoning? General truths are laid down as propositions to be proved, or in the forms of questions to be answered. Are these questions the child's or the teacher's? (Camp, 1903)

In an inquiry-oriented classroom, questions may be posed by the teacher *or* formulated by the children. The critical element to inquiry is that the child *seeks* answers to questions and is not *given* answers. "True learning comes from the search for the answer and not the answer—this is the essence of inquiry" (Renner, 1971).

Every description of inquiry includes the following: (1) active questioning and investigating, (2) acquiring new knowledge, and (3) observing and manipulating (mentally or physically) objects, phenomena, and/or nature. These characteristics of inquiry are present whether the investigative process takes minutes, days, or in many cases, years. Inquiry can be used by individuals not only in learning science but in thinking about many matters of interest in everyday life (AAAS, 1990). We have further defined inquiry within the teaching procedure known as the learning cycle. We will discuss the rational powers in the context of the learning cycle in Chapter 4.

Inquiry is a process that all individuals naturally use in approaching new situations and solving problems in life. By engaging in inquiry, the children in our schools gain experience with the mental activities that will improve their capacity to handle life situations and solve everyday problems. These mental activities include the use of the rational powers of the mind.

The Rational Powers

The rational powers of the mind seem to be woven into the process of logical thinking in a complex way, in order to perform their special function as needed. As mentioned, the rational powers include recalling, comparing, inferring, generalizing, deducing, classifying, analyzing, imagining, synthesizing, and evaluating.

Recalling Perhaps the most fundamental rational power is recall. To recall is to retrieve ideas that have been stored in the mind and bring them into conscious awareness. Without recall, no thinking is possible. Recall functions as the base of all logical thinking and production of information. The description of objects and events depends on the ability to recall properties used to describe and compare.

There are two reasons in the educational process for retrieving ideas. One reason is to pass the idea on to someone else—back to the teacher in many cases.

The other is to use the idea—to enhance it, relate it to other ideas, or gain new ideas through it.

If the primary use of recall in the education process is to retrieve ideas to pass on to someone else, then recall is used much like obtaining something from a grocery store in which ideas, like various products, are stored on shelves. This activity is simple recall. If the primary use of recall is to selectively retrieve ideas for use in thinking, then recall is used as a dynamic reservoir of ideas. This selective retrieval of ideas is enriched recall. To enrich recall, it must be used with other rational powers. In the educational process, recall is least likely to be enhanced through memorizing separate pieces of information to be repeated on cue. Recall is most likely to be enhanced by relating ideas about a subject to other ideas.

DO THE FOLLOWING ACTIVITY. IN FIGURE 2–1, WRITE YOUR NAME AND ADDRESS. BELOW YOUR NAME AND ADDRESS, WRITE YOUR TELEPHONE NUMBER, THE ALPHABET, A FRIEND'S NAME, AND THE TITLE OF THIS BOOK (WITHOUT LOOKING!). CAN YOU THINK OF OTHER INFORMATION THAT COULD BE COMMITTED TO SIMPLE RECALL? LIST THAT INFORMATION ON PAPER AND SHARE YOUR IDEAS WITH THE CLASS.

Name:	
Address:	
Telephone number:	
The alphabet:	
A friend's name:	
The title of this book:	

FIG. 2–1 *A simple recall activity*

You used simple recall in completing Figure 2–1. Simple recall is very important in everyday activities. But it should not be a goal of education. Activities designed to enhance simple recall are of little use in developing the logical thinking process. Also, substantive knowledge should never be taught as an exercise in simple recall—nor should evaluation focus on it. If large amounts of detailed or specific information are needed, then refer to an appropriate reference, for example, an encyclopedia. Books are written primarily to store and record large amounts of information for future reference. The mind is designed to *use* this information. Even if the information is stored in the mind, such vast information is difficult to "find" and retrieve when it is needed.

DO THE FOLLOWING ACTIVITY. IN FIGURE 2–2, DESCRIBE A BUTTERFLY. THEN, DESCRIBE YOUR BUTTERFLY IN FLIGHT, LANDING ON A FLOWER, AND TAKING TO FLIGHT AGAIN. DISCUSS YOUR BUTTERFLY DESCRIPTIONS WITH THE CLASS. HOW WERE YOUR DESCRIPTIONS SIMILAR TO THE DESCRIPTIONS GIVEN BY OTHERS IN THE CLASS? HOW WERE YOUR DESCRIPTIONS DIFFERENT? WHAT THINKING PROCESSES DO YOU THINK WERE USED IN "RECALLING" THE WAY A BUTTERFLY LOOKS IN FLIGHT AND UPON LANDING?

Description of a butterfly:
Description of the butterfly in flight:
Description of a butterfly landing on a flower:
Description of the butterfly taking to flight off of the flower:

FIG. 2–2 *An activity using enriched recall*

The activity in Figure 2–2 engaged you in the process of enriched recall. In contrast to simple recall, enriched recall requires the mind to use other rational powers in conjunction with recall. In this way, we capture information by using other powers along with recall, such as imagining, comparing, and/or analyzing.

Comparing Next to recall, comparing is the rational power most often used. It is very useful in producing information about an object or an event. Productive thinking does not take place with recall alone. Thinking and learning are extended through recalling and then comparing ideas that are known, according to similarities and differences. Changes in phenomena, for example, weather, are observed by using the rational power of comparing. Educational programs that focus entirely on simple recall are not contributing to students' ability to think. An educational program that links recall and comparing has taken the thinking process at least one step further.

IN FIGURE 2–3, DESCRIBE A BEE. DESCRIBE YOUR BEE IN FLIGHT, LANDING ON A FLOWER, AND TAKING TO FLIGHT AGAIN. COMPARE AND CONTRAST YOUR BUTTERFLY DESCRIPTION TO YOUR BEE DESCRIPTION. DISCUSS SIMILARITIES AND DIFFERENCES IN YOUR DESCRIPTIONS WITH THOSE OF OTHERS IN THE CLASS. WHAT MENTAL ACTIVITIES TOOK PLACE IN YOUR DISCUSSIONS WITH THE CLASS?

The activity in Figure 2–3 required the use of a number of rational powers including enriched recall and comparing. To make further use of these powers, one must go outside and observe butterflies and bees in nature. The rational power of comparing is enhanced by observing and recording similarities and differences between in-class activities and events or phenomena as they occur in nature.

Inferring In the two preceding activities you made comparisons, but you also used another rational power, inferring. This is the rational power used immediately after information is collected or confronted. This rational power has to do with the first seed of an idea regarding a pattern or trend; it is the rational power that initiates limited explanations that can later become more general and abstract. Inferring is the *explanation* of a single action, observation, or event in a larger collection of actions and events. Inference is the first small leap of the mind toward explaining data, and should be an integral part of almost every educational activity.

SELECT A PERSON TO STAND BEHIND THE DOOR AND HOLD ONLY ONE HAND AND ARM INSIDE. EXAMINE THE HAND. WHAT INFERENCES CAN YOU MAKE ABOUT THE PERSON FROM THE PERSON'S HAND?

In the last activity, notice that observations were made first, and inferences were formulated based on these observations. Observations are the foundation for

Description of a bee:
Description of the bee in flight:
Description of the bee landing on a flower:
Description of the bee taking to flight off of the flower:
Similarities between butterfly description in Figure 2-2 and bee description:
Differences between butterfly description in Figure 2-2 and bee description:

FIG. 2–3 *An activity using comparing*

making inferences. Making inferences requires more complex thinking than making observations. In making inferences we use logic and reason to explain our observations of objects, phenomena, and nature.

Generalizing Generalizing is an extension of our inferences. While an inference is limited in scope and usually applies to a specific action or event, a generalization takes into account a broader range of events and actions. For example, suppose a person has observed that a baseball bat, a log, and a table leg will burn. The person might infer that the objects burn because they are round and made of wood. Tests of other round objects, some of wood and others of glass or metal, would allow the person to conclude that indeed other round *wooden* objects do burn but round objects made of other materials do not.

If wooden objects with shapes other than round are investigated, they too are found to burn. The rational power of generalizing can now be used. The person can make the generalization, "all wooden objects burn." Notice that the generalization takes in the category of all objects made of wood, even those of shapes and kinds of wood that the investigator has not used. The generalization can be tested. That is, other objects made of wood can be tested to see if they actually do burn. But there still exists the possibility that an object made of wood might exist somewhere that will not burn. Testing can verify or increase the confidence in a generalization but does not *prove* it "true" even if the tests are positive. There is a close relationship between generalizing and hypothesizing. When engaged in an investigation, a researcher may propose several possible patterns or generalizations that could exist. These *tentative* generalizations guide further experimentation, which are usually referred to as hypotheses. The mental activity that takes place in generalizing also occurs when forming hypotheses. Thus, the rational power of generalizing *includes* hypothesizing.

Deducing Deducing is reversing the direction of one's thinking and using a generalization to predict or deduce a specific outcome or value. Inductive thinking or reasoning is reasoning from the specific (one idea) to a general rule. Deduction is using the rule that has been derived from inductive reasoning to find a specific idea.

WHAT ARE THE NEXT FOUR LETTERS IN THE FOLLOWING SEQUENCE: O, T, T, F, F, S?

The primary power that is used to solve the above puzzle is deducing. A pattern must be found in the sequence of letters, generalizations must made, and a solution must be deduced. The next four letters should be S, E, N, T, and the letters represent the first letters of the words one, two, three, four, five, six, seven, eight, nine, and ten. Note that individuals with previous experience with puzzles similar to the one given above may find it easier to decode the letters. These individuals were able to use generalizations from previous experience to make deductions about the nature of this problem, and about its solution. Thus, deducing is a powerful mental skill because it helps us make sense of new events and situations based on generalizations we had previously formed about similar events and situations.

Classifying Classifying is the systematic arrangement of objects, events, or ideas into groups or categories according to some established criteria. The rational power of classifying is one that often makes the use of other rational powers easier. If information has been classified or grouped, we need only recall the group, not each item. Once the group has been retrieved from memory, it is easier to recall the item. This activity is similar to locating an item in a file cabinet. Items of information are classified in folders according to similar content. To retrieve information about any item we retrieve the folder, and do not have to search through all of the unorganized information to find the item we need.

THINK ABOUT A DEPARTMENT STORE. IN FIGURE 2–4, RECONSTRUCT THE CLASSIFICATION SCHEME USED BY THE STORE. HOW DOES CLASSIFYING ITEMS MAKE IT RELATIVELY EASY TO FIND ITEMS IN A DEPARTMENT STORE? HOW ARE THE MENTAL ACTIVITIES INVOLVED IN CLASSIFYING IMPORTANT TO CHILDREN'S LEARNING IN SCHOOL?

Major groups of items	**Subgroup 1**	**Subgroup 2**	**Subgroup 3**	**Item**
EXAMPLE: Shoes	Children's shoes	Infants' shoes	Infants' casual shoes	Infants' casual shoes size 3 or size 3 sneakers

FIG. 2–4 *The rational power of classifying. To complete the figure, list major groups of items in a department store. Then list subgroups of the major items until individual items are listed (for example, shoes). A greater number of subgroups than provided in the figure may be necessary to classify certain items.*

In Figure 2–4, it is clear that items are easier to find and interpret when they have been classified. The same is true for children when they attempt to retrieve information they have stored in memory—it is easier for them to recall, infer, generalize, and make comparisons using this information. In other words, the rational power of classifying *facilitates* the use of other rational powers.

Analyzing To analyze is to separate and examine components. A sample of matter is analyzed to determine what substances are present and in what amounts. A speech or article is analyzed to determine what educational elements are present and the relative importance of each. To some extent, analysis also involves the idea of grouping or classifying, but the grouping is usually for the purpose of establishing relationships between or among categories.

EXAMINE A COLLECTION OF A VARIETY OF FLOWERING PLANTS. WHAT ARE THE PARTS OF EACH PLANT? WHICH PARTS ARE COMMON TO ALL OF THE PLANTS? WHAT DO YOU THINK ARE THE PURPOSES OR FUNCTIONS OF EACH PART? DO YOU THINK SIMILAR PLANT PARTS SERVE SIMILAR PURPOSES? WHY OR WHY NOT?

In completing the previous activity, you used the rational power of analyzing to mentally and/or physically separate whole plants into component parts. Analyzing was also used to determine common purposes or functions of these parts. Other rational powers were used in this activity, such as recalling, comparing, generalizing, and deducing. Analyzing is a fairly complex thought process because it incorporates the use of several other powers. Analyzing is also a crucial skill for developing higher reasoning and logic. Thus, it is important that children have experiences using the rational power of analyzing.

Imagining Imagining is the formation of a mental image of something that may not actually be present. Imagination is linked to the rational power of synthesizing to produce some of the most creative aspects of thinking—art, poetry, literature, music, architecture, computer technology and virtual reality, and model building in science. In order to develop this rational power, individuals must be engaged in creating new "things" or ideas using what is already known together with their imagination. Thus, children must be given opportunities to build models, write and tell stories, draw pictures, and represent ideas in a variety of ways. Science is a perfect medium through which children can use the rational power of imagination.

MAKE SOMETHING OUT OF PAPER THAT FLIES IN THE AIR A DISTANCE OF 20 METERS. DESIGN AND COLOR YOUR FLYING PAPER ANY WAY YOU WISH. DEVELOP, COORDINATE, AND CARRY OUT A FLYING PAPER "TEST FLIGHT" FOR EVERYONE IN THE CLASS TO FOLLOW. WRITE AN ORIGINAL STORY ABOUT YOUR FLYING PAPER. DRAW PICTURES WITHIN YOUR STORYBOOK. TELL YOUR STORY TO THE CLASS AND SHOW YOUR BOOK AND DRAWINGS. MAKE UP A POEM ABOUT YOUR EXPERIENCE WITH THE FLYING PAPER.

Notice that the previous activity did not use the word "airplane." This word was purposely omitted in order not to limit the imagination. Children often use their imaginations in play, and in making up games or playmates. So, not only does imagination develop thinking skills—children use it naturally, and they *enjoy* using it!

Synthesizing Synthesis is putting together parts to make a meaningful whole. The architect uses synthesis to create a building design. The cook uses synthesis to bake a casserole. The scientist uses synthesis to create a model that explains some aspect of nature.

Many drugs used to combat disease, fabrics from which clothes are made, and plastics serving various purposes were created by individuals using their imagination for the parts, and putting these parts together into a logical whole. Models, sketches, or computer animations of atoms and molecules were created through imagination and synthesis.

IN FIGURE 2–5, DRAW AN IMAGINARY LIVING ORGANISM. DETERMINE WHAT YOUR ORGANISM NEEDS OR USES TO SURVIVE AND REPRODUCE. USING THE ORGANISM'S REQUIREMENTS FOR SURVIVAL, CREATE A PLANET IN WHICH THIS SPECIES OF ORGANISM THRIVES. DESCRIBE ANY OTHER LIVING ORGANISMS YOU NEEDED TO CREATE TO SUPPORT THE LIFE OF YOUR ORIGINAL ORGANISM.

In Figure 2–5, imagination was the power used to create the organism and its individual needs for survival. Synthesis was used to put all of these imaginative ideas together to form the planet. In science, synthesis is frequently used to control variables and put together findings of laboratory activities in order to make sense of the experience. Synthesis is the power used to invent science concepts and/or formulate conclusions based on a combination of individual aspects of any laboratory or other learning activity. Synthesis is also used to link major science concepts together to better understand whole science disciplines (e.g., biology) and to better understand relationships between disciplines (e.g., biology and chemistry; biology and social studies).

Evaluating Evaluation is a comparison of something to some criteria. Evaluation involves making a judgment or forming an opinion about the worth of some condition, event, or observable situation. Educators often use evaluation to make judgments on the quality of students' work. The rational power of evaluation usually requires the use of all other rational powers in concert. To evaluate, one must recall facts, compare ideas, classify items and concepts, make inferences based on observations, analyze components, make generalizations and deductions, use imagination, and synthesize aspects of a situation. Students should be engaged in evaluation since it is considered the most complex and sophisticated form of thinking. Students need to evaluate their findings and conclusions following an exploration. Students must evaluate the salience of others' findings in light of their own and form arguments that support or refute such findings. Evaluation is what scientists do, and it is what contributing citizens of society do on a daily basis. Students must have experience with evaluating and forming opinions, grounded in evidence, to extend their thinking ability to a complex level.

Importantly, the children in our schools must use *all* of the rational powers to develop logical thinking skills that will empower them and maintain a free society for our future. It is the teacher's *obligation* to provide experiences that engage students in using powers of the mind that foster the development of logic and reason. Through these experiences, the teacher is making the learning of science

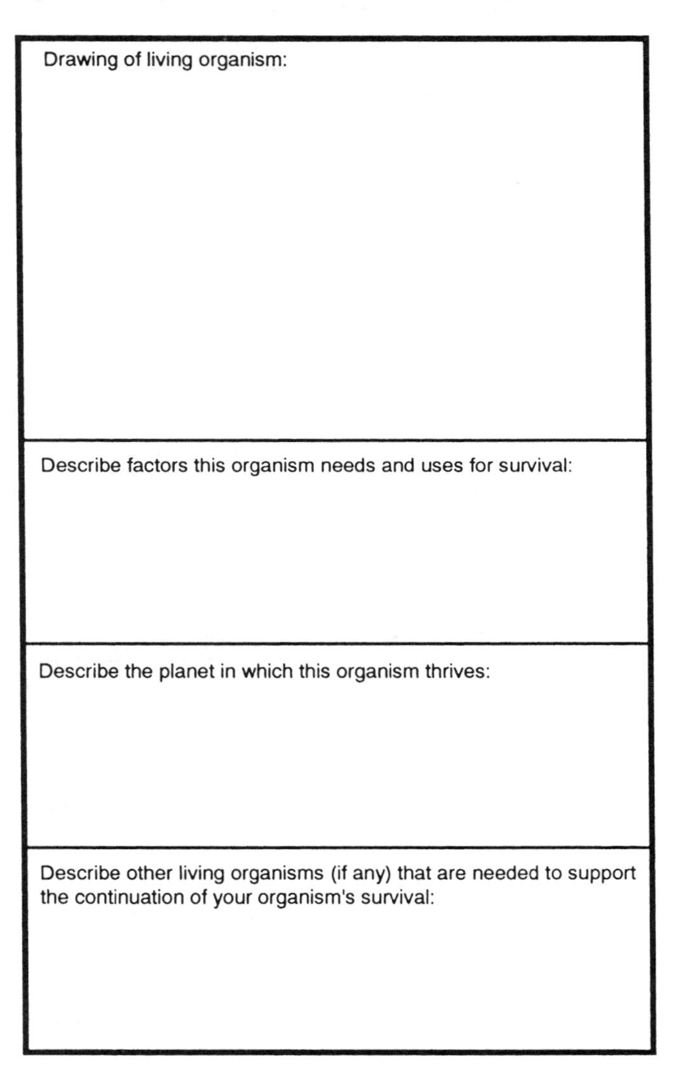
Drawing of living organism:

Describe factors this organism needs and uses for survival:

Describe the planet in which this organism thrives:

Describe other living organisms (if any) that are needed to support the continuation of your organism's survival:

FIG. 2–5 *Using synthesis to create living organisms and the environment in which they thrive*

content meaningful and relevant to students. The teacher is also preparing students to face problems in their lives every day, or problems that may face them in the future—and to face these problems with the rationality that leads to sensible and worthy resolutions. As a summary, Figure 2–6 shows science experiences teachers can provide for students that will incorporate the use of the rational powers and lead to the development of logical thinking ability.

Science Process Activities	Rational Powers Used
Collecting Data Observing Describing Experimenting	Comparing, Inferring, Recalling
Organizing Data Making tables Graphing Grouping Serial ordering Classifying	Classifying, Analyzing, Recalling
Interpreting Data Looking for relationships Constructing meaning	Inferring, Comparing, Recalling
Generalizing from Data Discerning a pattern Summarizing and proposing a trend Drawing a conclusion	Inferring, Generalizing, Synthesizing, Recalling
Explaining Generalizations from Data Making a model Creating or formulating a concept or idea Presenting data and conclusions to others	Imagining, Inferring, Recalling, Synthesizing, Evaluating
Predicting from Models or Patterns Deducing from a generalization Forming a hypothesis Testing a hypothesis, generalization or model	Deducing, Inferring, Recalling, Synthesizing, Evaluating

The Development of Logical Thinking

FIG. 2–6 *The rational powers used in science process activities to develop logical thinking (adapted from Renner, 1985)*

In Chapter 4 we will develop in detail the relationships between rational power development and the learning cycle. Take a few minutes to record your thoughts about the answer to this question: How do the phases of the learning cycle allow for the development of the rational powers of the mind? You may want to use Figure 2–6 to help you answer this question.

References

ALLING, M. 1881. "Natural Science in the Common Schools." *Education* (July): 601–615.

AMERICAN ASSOCIATION FOR THE ADVANCEMENT OF SCIENCE. 1990. *Project 2061: Science for All Americans*. New York: Oxford University Press.

CAMP, K. 1903. "Elementary Science Teaching in the Laboratory School." *The Elementary School Teacher* 3 (10): 661–667.

EDUCATIONAL POLICIES COMMISSION. 1961. *The Central Purpose of American Education*. Washington, DC: National Education Association of the United States.

EDUCATIONAL POLICIES COMMISSION. 1966. *Education and the Spirit of Science*. Washington, DC: National Education Association of the United States.

HUXLEY, T. [1869] 1964. "Scientific Education: Notes of an After Dinner Speech." *Science and Education*, eds. D. Runes and T. Kiernan. New York: Philosophical Library.

NATIONAL SCIENCE TEACHERS ASSOCIATION. 1994. *Handbook of Research on Science Teaching and Learning*, ed. D. Gabel, 51–70. New York: Macmillan.

RENNER, J. 1971. "Educational Purpose, Curriculum, and Methodology." *Journal of Thought* 6 (3): 162–167.

3 The Nature of the Learner

An adult and a five-year-old child are seated next to each other at a table. The adult carefully lays out a row of six red checkers and a second row of six black checkers. Opposite each red checker is a black one. The adult asks the child to count the number of red and black checkers. The child states that there are six red checkers and six black checkers. The adult asks, "Are there as many red checkers as there are black checkers?" The child says there are. The adult next says, "I am going to put the red checkers in a stack like this." The red checkers are now neatly stacked one on top of the other. "Now," says the adult, "do I have more red checkers, more black checkers, or is the number of the red checkers I have the same as the number of black checkers?" The child looks at the checkers and says, "There are more red checkers." "Why do you think so?" "Because the red checkers are taller." This child believes that just because the checkers are physically stacked there are now more.

Many times when those not familiar with children's reasoning observe a scene like this, they believe that the child did not understand the language. They think that if the adult had asked the "amount" of checkers rather than the "number" of checkers, the child would have reported correctly. If you feel this way, we urge you to get six black and six red checkers—or nickels and pennies or other similar objects—and a child between four and six years old and try the task just described. We have tried changing materials and language, putting the checkers in a pile as opposed to a stack, and anything else we could think of; the results were the same, although sometimes a child will say that there are more checkers in the row than in the stack.

When adults observe children using this kind of reasoning, they often conclude that young children do not reason. Quite evidently young children do reason; they reason differently from adults, but they do reason. Very frequently their reasoning leads them to conclusions that are at variance with the world, and then they—and possibly the world—become confused.

Why do children demonstrate the type of reasoning reflected in the examples above? What other characteristics do young children, and older children, display that can reveal why they reason as they do and what leads them to *construct* the understanding of the world they hold? The purpose of this chapter is to lead you to construct a model of how children learn and what they can learn.

We referred to leading children to *construct* their own understandings. What does that mean? We all "know" a lot of things, but much of what we "know" is not useful to us because it has no meaning. Asking children to memorize poems with abstract meanings for recitation to adult groups is an example of asking them to "know" something that usually has no meaning for them. No doubt we have all memorized something that had no meaning for us to get over some hurdle—a test, for example. *Such material is not knowledge; knowledge must have meaning.*

When we interact with a new object, event, or situation, we participate in all aspects that are available to us. We select what *we* believe are the salient facts, ideas, and relationships from that interaction and put them all together into a whole that is meaningful to us. The complexity of that meaningful whole obviously changes with our experiences; this is the point of the reasoning model discussed here. We develop knowledge from the kind of interaction just described. Such knowledge is usable with new objects, events, or situations because we, ourselves, built or *constructed* our understanding from our interactions. Our meaningful understanding is ours and not someone else's, about which we were told. We can have assistance in constructing our knowledge about something, but if we are permitted, we will construct our own knowledge of the object, event, or situation. Even if we are *instructed* in exactly what we should know and how we should know it, we probably still construct our own understanding, because most of what we memorize, we forget. What is left is what we have constructed for ourselves. We believe that every student in school should have the opportunity to engage in *knowledge construction*; this is what everyone does outside the classroom and in adulthood.

When considering children's learning, there is more than one model that can be used. This book, in general, is based upon a constructivist model for learning. That model is concerned with the mental processes involved in knowing: perception, imagery, reasoning, and so on. Central to this book is the developmental theory of the Swiss psychologist and epistemologist Jean Piaget. He and his associates have been studying the development of cognitive processes in children (and publishing their findings) since the 1920s. In this chapter we will cite from Piaget's and his associates' seminal references. Keep in mind that Piaget's theory is basically cognitive and is essentially a model of intelligence. In fact, Piaget (1969) has stated that the basic aim of his work has been "to explain the development of intelligence and to comprehend how from elementary forms of cognition superior levels of intelligence and scientific thinking came about." Frequently, it has been said that the theory of Piaget is a model of intellectual development, which it is. But do not confuse the intelligence that is Piaget's concern with that which is measured by an IQ test. What Piaget has in mind is intelligence that directs our interaction with our surroundings and the persons and things in it. He is concerned with the development of the entire intellect.

Actions and Operations

> Knowledge is not a copy of reality. To know an object, to know an event, is not simply to look at it and make a mental copy or image of it. To know an object is to act upon it. To know is to modify, to transform the object, and to understand the way the object is constructed. An operation is thus the essence of knowledge; it is an interiorized action. (Piaget, 1964a)

Piaget states that to know an object or event, the learner must act upon it; there are, however, different ways of acting. Very young children simply hit, grab, squeeze, and do all manner of physical things in interacting with an object. Their actions are entirely overt. Next comes a time when "the young child simply runs off reality sequences in his head just as he might do in overt action" (Flavell, 1963). The child has made, according to Flavell, a step-by-step mental replica of concrete actions and events. In other words, the child who is capable of actions with objects and events is interacting by running off mental "reality sequences" with them; that child is really trying to reproduce reality.

Eventually the child has had enough experiences, and a dramatic event occurs. Piaget explains it in this way:

> When he [the child in the example] was . . . a small child—he was seated on the ground in his garden and he was counting pebbles. Now to count these pebbles he put them in a row and he counted them one, two, three, up to ten. Then he finished counting them and started to count them in the other direction. He began by the end and once again he found ten. He found this marvelous that there were ten in one direction and ten in the other direction. So he put them in a circle and counted them that way and found ten once again. Then he counted them in the other direction and found ten once more. So he put them in some other arrangement and kept counting them and kept finding ten. There was the discovery that he made.
>
> Now what indeed did he discover? He did not discover a property of pebbles; he discovered a property of the action—which he introduced among the pebbles. The subsequent deduction will consist of interiorizing these actions and then combining them without needing any pebbles. (Piaget, 1964a)

The child in the example made a dramatic move, from creating a step-by-step mental replica of reality to mentally transforming and modifying what reality is. The child began to do what Piaget calls *mental operations*; the child began to construct knowledge. The primary difference between actions and mental operations is that actions mentally reproduce reality, while operations do something further with that reproduction.

Consider the example cited earlier of the child and the two lines of checkers. The child who believes that there are more checkers in the stack is simply making a mental replica of the height of the stack compared to the length of the line of checkers. That child cannot reverse the thinking process to include the fact that the line is also much shorter than the stack. All the child can do is make a mental replica of the reality of both objects. But the child who can do mental operations can mentally reverse the stack-of-checkers image to the line-of-checkers image. This learner can indeed transform the data received from the environment; the learner has interiorized the action of producing a stack from the line. Piaget describes a mental operation as "*an interiorized action.* But in addition, it is a *reversible action*; that is, it can take place in both directions, for instance, adding or subtracting, joining or separating" (Piaget, 1964a). Think of mental operation this way: it is *mentally* doing something with the data received from the environment.

A CHILD PLANTS A SEED IN A CUP AND TAPES ANOTHER SEED TO THE OUTSIDE. WHEN THE PLANT APPEARS IN THE CUP THE CHILD WHO PLANTED IT DOES NOT SEE THE CONNECTION BETWEEN THE SEED ON THE CUP AND THE PLANT. DEFEND WHETHER THIS IS EVIDENCE OF AN ACTION OR A MENTAL OPERATION.

A Model for Levels of Intellectual Development

In 1920 Piaget accepted a position in the Binet Laboratory in Paris. His assignment was to develop standardized French versions of certain English reasoning tests. While carrying out this responsibility he made two major findings that led him into the detailed study of intelligence. He found that children of about the same age frequently gave the *same wrong answers* to a particular question. Additionally, at different ages there were different kinds of common wrong answers. These findings led him to believe that the thinking of younger children was *qualitatively* different from that of older children. In other words, younger children actually believe that the world works in a way different from what older children believe.

These findings led Piaget to reject a quantitative definition of intelligence, which is derived from measuring the number of correct responses on an IQ test. According to Piaget's theory, a young child may be just as bright as an older one, but the qualities of the types of thought of which the two groups are capable are distinctly different. Adults often believe that older children are brighter than young children because the thought type of older children more closely approximates adult thought. This notion is built around the assumption that children are miniature adults. The Piagetian model rejected this idea and stated that a certain age range has a distinct quality of thought. According to the Piagetian model of intellectual development, humans pass through stages or phases of thought in moving though life. Thought in each stage has certain properties that differ from those in other stages.

Many years of exacting study of children at all ages have gone into Piaget's model of stages of intellectual development. Piaget's data about the reasoning patterns of humans begin at birth and extend into the third decade of life. Obviously, the data gathered from birth and for several years thereafter are based only on Piaget's observations.

Perhaps, for our purposes, Piaget's procedure for gathering data could be described as giving the child a task that involves material objects and reasoning, letting the child perform the task, and then asking the child what she did and why she did it that way. What is important for you as a teacher is that *Piaget's model of intellectual development is derived from direct association with learners of all ages*. Any model used by teachers to guide them in selecting and employing content and materials must be relevant to children.

The data that Piaget and his coworkers gathered led to the formulation of a model of intellectual development that includes four unique levels. The quality of thought in each of these levels (stages, phases, periods) is distinctly different from the quality of thought in each of the other levels. In other words, the content children at each level can learn is unique. Do not interpret this statement to mean that all children in a particular level think *exactly* alike, but rather, that the thinking of children in the same level has common properties.

The First Level

The first stage of intellectual development in Piaget's model begins at birth and continues for approximately two and a half years. Piaget has called this period the *sensorimotor stage*. During this stage the child learns that objects are permanent—just because an object disappears from sight does not mean that it no longer exists. Acquiring the characteristics of object permanence explains, for example, why a child approximately a year old will cry when the mother leaves. This separation anxiety, however, does not occur earlier because until that point, "out of sight, out of mind" adequately describes the child's perception. During the sensorimotor period, language begins to develop—a development that is far too complex to explore fully in this book. Basically, however, the child learns how to attach sounds to the objects, symbols, and experiences he has had. But inventing appropriate sounds for something depends, as does later learning, on the child's experiencing that something.

During the sensorimotor period, the first signs begin to emerge showing that intellect is *developed* and not spontaneous. Now, certainly, the way a child in the sensorimotor stage goes about learning is quite different from the way an adult learns. But throughout all the stages of Piaget's model, the fact becomes obvious that later learning cannot occur unless early learning has been accomplished. This means that for culturally deprived children who have not had the benefit of a rich environment to assist them in developing the beginnings of a language system, the school may need to provide many experiences that go far beyond conventional reading readiness programs before traditional school activities can begin.

There is little likelihood that you will be working with children in the sensorimotor stage. You need to be aware, however, that this is the stage in which intellectual development begins to emerge, and until the child accomplishes certain goals in this stage, later learning must wait. *Perhaps we, as teachers, need to spend more time determining what the learner is ready to learn and less time being concerned with the specific content being covered.*

Much confusion has developed about the ages at which children move from stage to stage within the intellectual development model. Piaget has repeatedly said that the ages he suggested are only approximations and has gone so far as to say that to "divide developmental continuity into stages recognizable by some set of external criteria is not the most profitable of occupations" (Piaget, 1963). The external criterion most often used is chronological age, but using it can be misleading.

There is only one stage in Piaget's model whose starting point can be precisely stated—the sensorimotor stage. As noted earlier, it begins at birth and ends around two and a half years of age. A two-and-a-half-year-old child will begin to enter the *preoperational stage*; the exit from this type of thinking begins around seven years of age. In other words, the model does not state precise ages at which a learner will progress from stage to stage. Piaget explains that, although the order of succession is constant, the chronological ages of these stages varies a great deal (Piaget, 1964a). As you read the remainder of this discussion about Piaget's model of intellectual development, keep in mind that a child does not move completely from one stage to another at one time. The evidence available suggests that a learner can easily be in the sensorimotor phase on some traits and in the preoperational phase on others. Rather than thinking that a child moves from one stage to another, think of the child as developing certain traits of a particular stage. As development progresses, the child moves more deeply into one stage on some traits and begins to move into the next, higher stage on other traits. In other words, there is not a chronological line children cross indicating that they have moved completely from one stage to another, like the line crossed when an individual is permitted to vote at age eighteen.

IN THIS PARAGRAPH THE NOTION OF "TRAIT" IS MENTIONED. WHAT IS A TRAIT? BASED ONLY ON WHAT YOU KNOW SO FAR, DESCRIBE WHAT YOU THINK ARE THE "TRAITS" MENTIONED.

The Second Level

Think back to the discussion regarding the difference between "action" and "operation." The name of the second stage in the Piagetian mode, *preoperational*, is wonderfully descriptive of what the mental activities of children occupying that stage are like. Children of this age are confined to making step-by-step mental replicas of the environment and running off reality sequences in their heads. However, doing something with those reality sequences—that is, performing mental operations—cannot be accomplished by the preoperational child. Preoperational children see, decide, and report. They think, but thinking *about* what they think is beyond their intellectual ability. The *pre*operational stage of thought is one that exists *before* mental operations are possible.

A complete description of all the intellectual characteristics of the preoperational child is far beyond the scope of this book. If you wish to investigate the characteristics of preoperational children further, we suggest that you consult Piaget's *Psychology of Intelligence* (1963), in which he explains his intelligence model and the characteristics of the stages within the model.

As children begin to develop preoperational thought, certain changes occur in their cognitive abilities. Piaget and Inhelder (1969) explain these changes in this way:

> At the end of the sensori-motor period . . . there appears a function that is fundamental to the development of later behavior patterns. It consists in the ability to represent something by means of a "signifier" . . . which serves only a representative purpose. . . . We generally refer to this function that gives rise to representation as "symbolic." However, since linguists distinguish between "symbols and signs," we would do better to adopt their term "semiotic function" to designate those activities having to do with . . . signifiers as a whole.

This quotation describes an important difference between the sensorimotor child and the preoperational child. Carefully consider the phrase "which serves only a representative purpose." This is what is meant by a "signifier"; it represents what is going to happen. Thus, when the preoperational child hears the bell of the ice cream truck, it signifies to the child that ice cream is available. To the sensorimotor child, however, the sound of the bell and the taste of ice cream are not mentally separated. Hearing the sound of the bell also means the taste of the ice cream. A sensorimotor child sees a sibling put on a coat and thinks the sibling is going to school. In other words, the overt sign—the signifier—is not differentiated from what the sign means.

A preoperational child can think about signifiers and distinguish them from the objects, events, or situations they signify. This ability allows the teacher to use some definite characteristics in working with preoperational children. There are at least five of these characteristics, called semiotic (or symbolic) functions, whose appearance is almost simultaneous. Each of the five is described, from the simplest to the most complex, in the following discussion.

1. Deferred imitation: This is the ability of a child to observe an event, object, or situation and later imitate what has been observed. Children who have observed how a rabbit wrinkles its nose and later try to do it when asked are using this preoperational characteristic. The establishment of habits uses a good bit of deferred imitation.

2. Symbolic play: This trait is most adequately described by the preoperational child's game of pretending, a game that is not found at the sensorimotor level.

Using this trait allows a teacher to have a play store in the classroom, for example.

3. Drawing: Here the child is able to represent experiences graphically. Many times the symbols used are not clear to an adult, but the child can explain them. Piaget believes that this trait is an intermediate stage between play and mental images. If you doubt the importance of mental images to thinking, try to imagine thinking without them. (That, too, is a mental image!)

4. Mental image: This is, of course, what must be available to children before they can describe anything they have experienced. Whenever a teacher says, "Tell me what happened" or "Tell me what you saw," the ability to use the mental-image trait is assumed.

5. Verbal evocation: The increasing ability of children to use language makes it possible for them to describe events that have already happened.

When children describe or comment on something that has recently happened, they are using at least the mental images and verbal evocation traits. Obviously, the five semiotic functions are interconnected.

Why is the semiotic function of young children important for their teachers to know? Answering this question requires the evolution of a principle for using the Piagetian model in the classroom. If children are to develop increasingly complex mental abilities and content understanding (undergo intellectual development), they must have maximum opportunities to interact with their environment. Preoperational children *can* use symbolic play, drawing, and all the other semiotic functions. Instruction, therefore, must also use these functions.

The teaching principle that can be deduced from Piaget's model says that teaching at any level of intellectual development must use the mental abilities unique to that particular level. If teachers insist that children "know" something when they do not have the mental abilities to understand what is being pushed at them, children have only one choice—they memorize. When the trivia contest is over, most of what they have memorized they promptly forget. We will examine this point further as we look at the next levels in Piaget's model.

DESCRIBE A CLASSROOM ACTIVITY THAT WOULD USE EACH OF THE FIVE SEMIOTIC FUNCTION TRAITS.

For the purposes of applying the Piagetian model in selecting and using content and instructional methodology, six basic characteristics of the preoperational child—in addition to the semiotic function—warrant examination. The characteristics are the following:

1. egocentrism

2. irreversibility

3. centering

4. states in a transformation

5. transductive reasoning

6. conservation reasoning

Egocentrism is one of the preoperational child's most prominent traits. The child sees the world from only one point of view—her own. The world revolves around the child, who is completely unaware of being limited to only one way of viewing it. The preoperational child cannot see another's point of view and coordinate it with his own and that of others. Children's perceived opinion—a reality sequence—is what they believe, and they feel no responsibility to justify a belief or to look for contradictions in it. Preoperational learners have developed a certain language pattern that they use to communicate with others, *but* they do not have the ability or see the need to adapt that language pattern to the needs of their listener.

The second trait of the preoperational child, which has great importance to curriculum and teaching methodology, is *irreversibility*. In order for humans to begin to perform intellectual operations, they must be able to reverse their thinking. Irreversibility of thought is beautifully illustrated by this dialogue with an eight-year-old boy (Piaget, 1964b):

Have you got a brother?
Yes.
And your brother, has he got a brother?
No.
Are you sure?
Yes.
And has your sister got a brother?
No.
You have a sister?
Yes.
And she has a brother?
Yes.
How many?
No, she hasn't got any.
Is your brother also your sister's brother?
No.
And has your brother got a sister?
No.

The dialogue continues until finally the child recognizes that he is his brother's brother. This dialogue with a four-year-old girl also nicely illustrates the irreversibility concept (Piaget, 1964b, p. 85):

Have you got a sister?
Yes.
And has she got a sister?
No, she hasn't got a sister. I am my sister.

Reversibility means that a thought is capable of being returned to its starting point. For example: $8 + 6 = 14$ and $14 - 6 = 8$. The thought started with eight and returned to eight. Preoperational children cannot reverse their thinking. Consider what this says to those planning a mathematics program for the early primary grades. Much of our society has a real "hang-up" about mathematics. Perhaps understanding many of the mathematics concepts presently taught in an early elementary mathematics program requires mental reversibility. Since children cannot use reversibility, they cannot develop the understanding demanded. They have no choice, therefore, but to use rote memory if they wish to survive the tests. Could it be that such an experience creates hang-ups about mathematics that individuals never conquer?

Isolating the irreversible trait in a young child's thinking is not difficult, and it is informative. A procedure using clay will enable you to do it. The materials you will need are simple—two equal quantities of modeling clay or plasticene (we found that using different-colored pieces facilitates communication with the child).

FORM THE PIECES OF CLAY INTO TWO BALLS AND EXPLAIN TO THE CHILD THAT YOU WANT TO START THE EXPERIMENT WITH ONE BALL JUST THE SAME SIZE AS THE OTHER. IN OTHER WORDS ONE BALL SHOULD CONTAIN *JUST AS MUCH* CLAY AS THE OTHER. ALLOW THE CHILD TO WORK WITH THE TWO BALLS UNTIL *THE CHILD BELIEVES* THE BALLS CONTAIN THE SAME AMOUNTS OF CLAY. NOW, DEFORM ONE OF THE BALLS; A GOOD WAY TO DO THIS IS TO ROLL ONE OF THE BALLS INTO A LONG, CYLINDRICAL SHAPE OR A PANCAKE.

NEXT, ASK SEVERAL CHILDREN (FIVE-YEAR-OLD CHILDREN ARE PROBABLY BEST) IF THERE IS MORE CLAY IN THE BALL, MORE IN THE ROLL, OR THE SAME AMOUNT IN EACH (BE SURE TO GIVE EACH CHILD ALL THREE CHOICES). ASK EACH CHILD WHY THE ANSWER GIVEN IS THE ONE BELIEVED CORRECT. RECORD THE RESPONSES YOU RECEIVE.

A child who has not developed the thinking trait of reversibility will tell you that there are different amounts of clay in the two shapes. Our experience has been that most preoperational children will select the cylinder (or pancake) shape as the one containing more clay.

The preoperational subject is not able to make the thinking reversal from the cylinder-shaped object that now exists back to the clay sphere that did exist. The child cannot do the analyzing and synthesizing that would permit her to reconstruct the sphere mentally, although the child knows that the sphere existed. This can be proved by asking the child to restore the roll of clay to its original shape; a sphere will be produced and the child will now tell you that there is the same amount of clay in each. Children of this age think, but their thinking is so irreversible, they cannot think *about* their thinking.

Why does the preoperational learner usually focus attention on the cylinder-shaped object rather than on the ball? This can be explained by using another trait in the preoperational model—*centering*. When one clay ball was deformed, the child's attention was probably fixed on the detail of length, and rigid, preception-bound mental abilities prevented the child from seeing anything else about the transformed object. Educational experiences provided for young children must avoid using materials, activities, or both that encourage the centering trait. If colors are used, for example, they should all be attractive and appealing. Teachers must not be surprised when children focus their attention on one aspect of an object, event, or situation; they are only acting as preoperational learners can be expected to act.

Centering is a characteristic of preoperational children, and those working with them should expect to find it. Does a child's inability to reverse thinking cause centering, or does the child's centering trait cause irreversibility? Who knows? Besides, is it important? Both traits exist and which comes first is really not relevant because obviously they are not mutually exclusive.

The extreme perception-boundness of a preoperational child is well illustrated by the trait known as *states in a transformation*. Figure 3–1 represents a wooden rod that is held vertically (position 1) and then released (positions 2, 3, 4, and 5). The rod eventually comes to rest at position 6. Obviously, the rod was in a state of rest when it was held in position 1 and is again at rest in position 6. If a series of pictures is taken of the falling object, it would be seen to pass through many other states, represented by positions 2, 3, 4, and 5. In other words, the series of states in the event results in a transformation, from the stick standing erect to the stick lying in a horizontal position.

If preoperational children are shown the experiment, after having been informed that they will be asked to draw a diagram of it, they will not draw what is shown in Figure 3–1, nor will they indicate in any way what successive states the stick went through in being transformed from position 1 to position 6. Our experience in asking children to do this task has been that they draw only positions 1 and 6. They see only the beginning and final states and do not see the transformation. This particular preoperational trait (which also shows irreversibility and centering) is particularly important when young

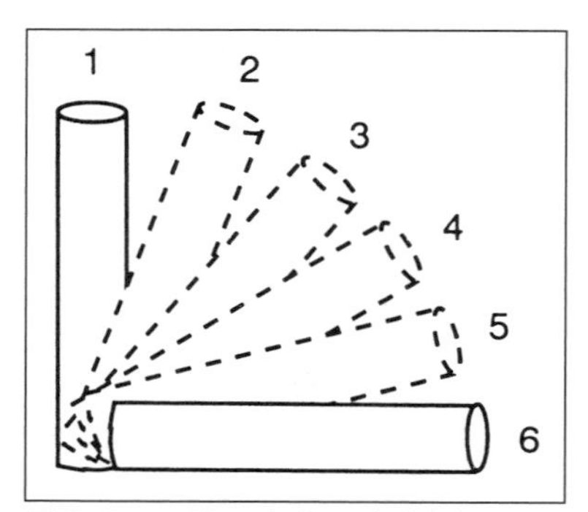

FIG. 3–1 *The ability of children to see states in a transformation can be assessed with the falling rod experiment.*

children are studying science and doing experiments, as, for example, a plant-growing experiment. There is little need to try to get them to see the importance of the several states in the transformation; *they cannot do it*. They will perceive the first and final states and nothing else. The process that allowed the final state to be a function of the intermediate states cannot be understood by preoperational children. This statement seems to call into doubt the insistence that young children do *detail* experiments; they will see the beginning and the end, but as long as they are preoperational, they will not learn the transformation of objects through the process of experimentation.

At the beginning of this chapter you read of an incident in which a young child made decisions using peculiar reasoning: that stacking up a row of checkers increased their number. Such reasoning seems strange, even silly, to an adult, but it is very real to a child. This child is not reasoning from a general to a particular case (deduction) nor from a series of particular cases to a generalization (induction). Rather, the child is reasoning from one particular case to another particular case through *transductive* reasoning. This type of reasoning begins to appear in the child with the beginning of language and generally lasts until after four years of age.

In our study of children, we have often used Piaget's classification tasks. One of these involves showing a child a great number of wooden beads, all of one color (we usually use red), and a few of another color (say, blue). Ask the child if there are more wooden beads or red beads. A truly preoperational child will tell you there are more red beads, and when asked why, will often answer, "Because they are prettier"—which is a *transductive* response. As a teacher, do not be surprised if you encounter such transduction in kindergarten and first-grade children. If you do, be patient; usually it disappears with increased experiences in the school environment.

REREAD THE CHECKERS EXAMPLE AND WRITE A COMPLETE EXPLANATION OF WHY THAT CHILD IS USING TRANSDUCTIVE REASONING.

Identifying whether or not the preoperational thinker can see the relationship between states in a transformation is a simple task; you do the falling stick experiment with the children and then ask them to tell you what happened. You have just explained why the checkers activity allows a child to demonstrate transductive reasoning, and other tasks will also do it. Identifying egocentrism, irreversibility, and centering, however, is not as easy as using the falling stick experiment.

There are procedures, however, that can be used to identify the traits of preoperational children. You have already met two of the techniques used—the activities involving checkers and clay balls. Those activities illustrate the inability of a preoperational child to mentally hold the image of an object and to see that distorting the object does not change the amount of material it contains. Mentally holding the original image of an object is called *conservation reasoning*. Preoperational children do not conserve; they make decisions about the distortion of the object on the basis of what they perceive. This rigid perception-boundness, however, is due to these children's irreversible thinking, tendency to center, extreme egocentrism, inability to see a transformation between states, and transductive reasoning. Isolating a child who does not use conservation reasoning will allow you to describe that child's stage of intellectual development in terms of the preoperational traits we have already described *and* the trait of conservation reasoning. According to Piaget, "the clearest indication of the existence of a preoperational period . . . is the absence of notions of conservation until about the age of seven or eight" (Piaget and Inhelder, 1969).

Conservation, then, is an overt manifestation of whether or not a child is a preoperational thinker. As we said earlier, this stage of development begins at about two and a half years of age. In describing the beginning of a child's ability to conserve, Piaget has also provided information about the end of the preoperational period:

> There always comes a time (between 6 and one-half years and 7 years 8 months) when a child's attitude changes: he no longer needs to reflect, he decides, he even looks surprised when the question is asked, he is certain of the conservation. (Piaget, 1963)

The beginning of the ability to conserve and the beginning of the child's entry into the third stage in the Piagetian model—*concrete operations*—occur, then, in the late first or early second grade. For purposes of designing a first-grade curriculum for most of the year, a teacher can consider that the children are preoperational.

The clay balls experiment measures the ability to conserve solid amount, and the checkers experiment assesses whether or not the child can conserve number. There are four additional tasks we have found useful. These tasks are the conservation of liquid amount, length, weight, and area. Descriptions of all six tasks and how they are administered follow. As you read these descriptions, keep in mind the definition of "conservation," which may be stated thus: Children who conserve can hold a concept about an object in their minds while a second object, which is like the first, is distorted, and they can see that the distorted object is still like the nondistorted object in many specific ways.

> CONSERVATION OF NUMBER TASK. You have already met this task. Have the children line up six black checkers in one row and six red checkers in another row, as shown in Figure 3–2A. Ask the child if he agrees that there are as many red checkers as there are black checkers. After the child agrees, stack the red checkers, one on top of the other, and leave the black checkers as they were; the checkers will now appear as in Figure 3–2B. After you have rearranged the checkers, ask if there are more red checkers, more black checkers, or if the numbers of black and red checkers are the same. If the child reports that the numbers are the same, number is conserved. *Be sure to ask why the child believes as he does, not*

only on this task but also on all the others. Getting the child to explain his answer will tell you a great deal about the state of the child's intellectual development.

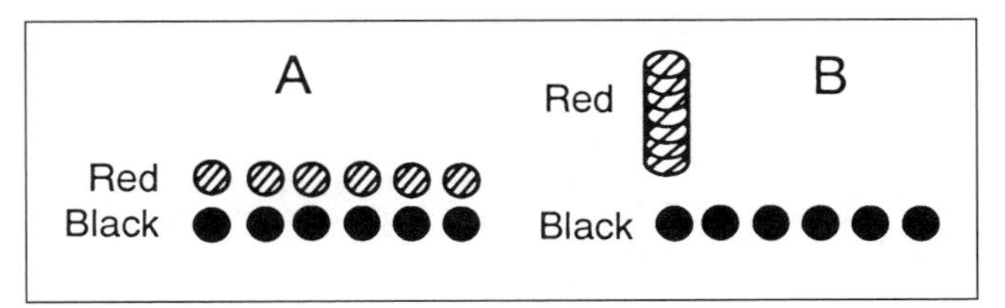

FIG. 3–2 *Conservation of number task*

CONSERVATION OF LIQUID AMOUNT TASK. Pour the same amount of water into two containers of equal size (see Figure 3–3A). For convenience, you may wish to color the water in one container. Ask if the child agrees that the containers are the same size and that they contain the same amount of liquid; if the child wishes to adjust the water levels, let this be done. After agreement has been reached that the amounts are equal, have the child pour one of the liquids into a taller, thinner container (see Figure 3–3B) and ask if there is more colored water, more clear water, or if the amounts are equal. A report that the amounts are equal shows that the child conserves liquid amount; a report that there is more water in one of the containers demonstrates a lack of liquid-conservation ability.

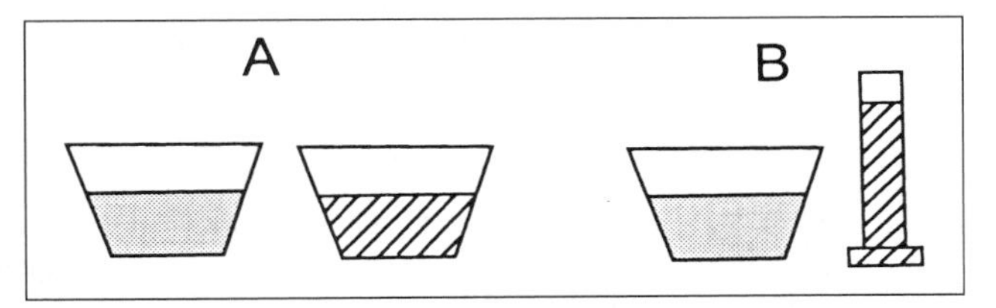

FIG. 3–3 *Conservation of liquid amount task*

CONSERVATION OF SOLID AMOUNT TASK. This task has already been cited. Prepare two pieces of clay, each containing the same amount, and roll them into balls of equal size (see Figure 3–4A). For convenience during the discussion with the child, you may wish to use two colors of clay, blue and red, for example. Ask the child if there is the same amount of blue clay as red clay; let the child make any adjustments in the clay balls she feels are necessary. Next, deform the piece of red clay by rolling it into what you may want to call a "snake" (see Figure 3–4). Ask the learner if there is more clay in the ball, in the snake, or if there is the same amount in each. Recognizing that the amount of the solid remains constant indicates solid-amount conservation ability.

CONSERVATION OF AREA TASK. Begin this task by showing the child two pieces of green construction paper of *exactly* the same size. Explain that the pieces of paper represent two fields of grass and that they must be exactly the same size. Give the child the opportunity to examine the two pieces of paper and to make

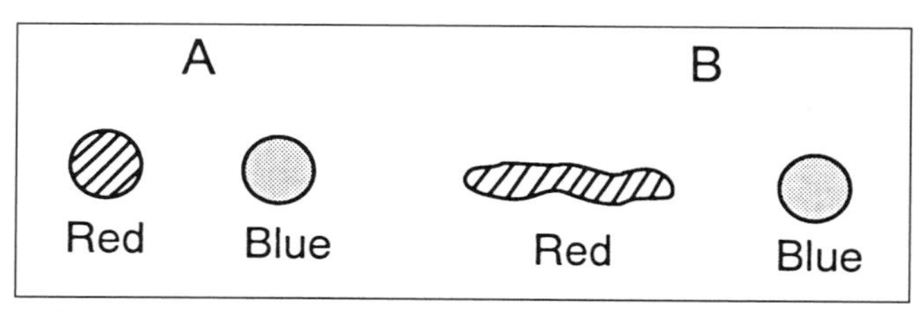

FIG. 3–4 *Conservation of solid amount task*

adjustments on them if necessary. The child must be convinced that the two fields are exactly the same size. Be sure to ask the child to explain to you why one field contains as much grass as the other. That act of explaining convinces the child to think about the papers as grass (a preoperational child can do this; think about the semiotic functions) and will indicate that the child understands the "sameness" concept in this task.

Next show the child some wooden cubes (about 2.5 centimeters on a side). Explain that you are going to pretend that each of these cubes is a barn. Give the child the opportunity to examine the "barns" until the child is convinced they are all the same size. Place a barn on each field (see Figure 3–5A) and ask the child whether, in each field, there is still the same amount of grass that has not been covered, or whether one field or the other has more grass. After adding a barn to each field and having the child tell you which has more uncovered grass, *always* ask why the child believes the answer given is true. Record the child's answer. Now place one more barn on each of the fields, but on one field of paper place the barns very close together and, on the other, separate them (see Figure 3–5B). Again ask the child if there is still the same amount of uncovered grass in each field and why. Use a third barn as shown in Figure 3–5C and again ask questions about the amount of uncovered grass.

Not infrequently, the only justification a child gives to explain why the grass area is the same in both fields in that the numbers of barns on the two fields are equal. These data suggest that the child is conserving number and not area. To determine whether or not this is what the child is doing, follow the procedure shown in Figure 3–5D; here one of the barns has been placed on top of another. Ask the child the questions as before. The child who responds by stating that the amounts of uncovered grass are equal in Figures 3–5A, B, and C but unequal in Figure 3–5D has demonstrated the ability to conserve area.

CONSERVATION OF LENGTH TASK. This task requires a wooden dowel forty centimeters long and four dowels of the same diameter each ten centimeters long. Straws substitute well for wooden dowels. The exact lengths and numbers of the shorter dowels are not important, but the combined lengths of the smaller dowels must equal the length of the long dowel. Two identical toy cars are also helpful. Place the long dowel and the shorter pieces parallel, so that the combined length of the pieces just equals the length of the long piece (see Figure 3–6A). Be sure the child agrees that the line of pieces is exactly the length of the long piece; let adjustments be made if necessary. Inform the child that the dowels represent roads and there is going to be a race. Place identical toy cars

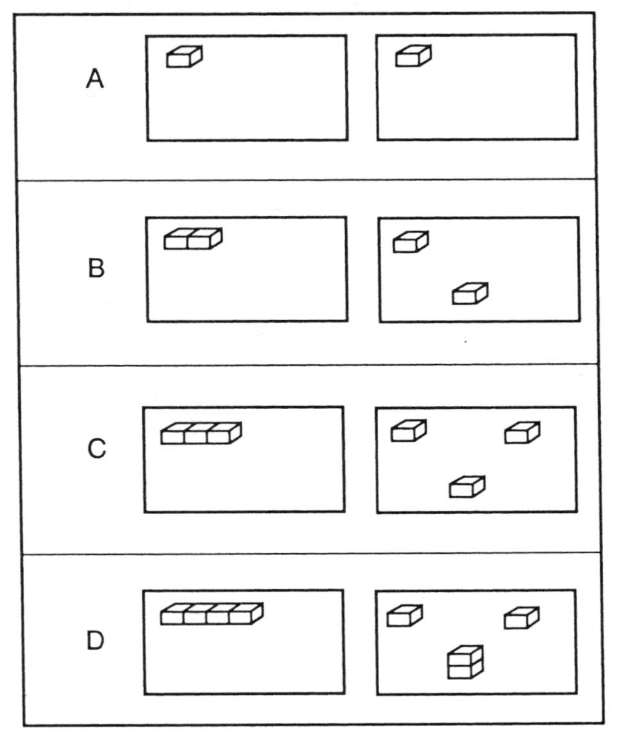

FIG. 3–5 *Conservation of area task*

(say, a red car and a blue car) at the same ends of the roads and then pose this question: "If the cars travel at the same speed, which car, the red one or the blue one, will reach the end of the road first? Or will they reach the ends of the roads at the same time?" If the child does not ultimately agree that the cars will reach the ends of the roads at the same time, abandon the task.

Next, move two pieces of the four-piece road as shown in Figure 3–6B and ask the questions about the race. If the child states that the cars will reach the ends of the roads at the same time and also gives the correct reason, he conserves length.

CONSERVATION OF WEIGHT TASK. Give the child two balls containing equal weights of clay; two colors of clay, such as red and green, facilitate communication in this task (see Figure 3–7). Add and subtract clay from each of the balls until the child agrees that the balls weigh exactly the same. Next, take the two balls of clay from the child and flatten one of them into a pancake or distort it in some other way. *Don't let the child lift the two clay objects after this distortion.* Next, ask the learner if the green clay weighs more, the red clay weighs more, or if the weights are still the same. Failing to recognize that the weights of the red and green clay objects are still equal shows that the child does not conserve weight.

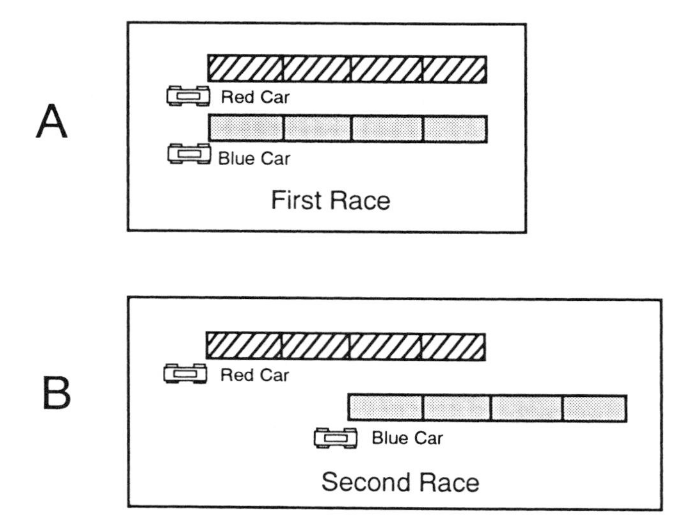

FIG. 3–6 *Conservation of length task*

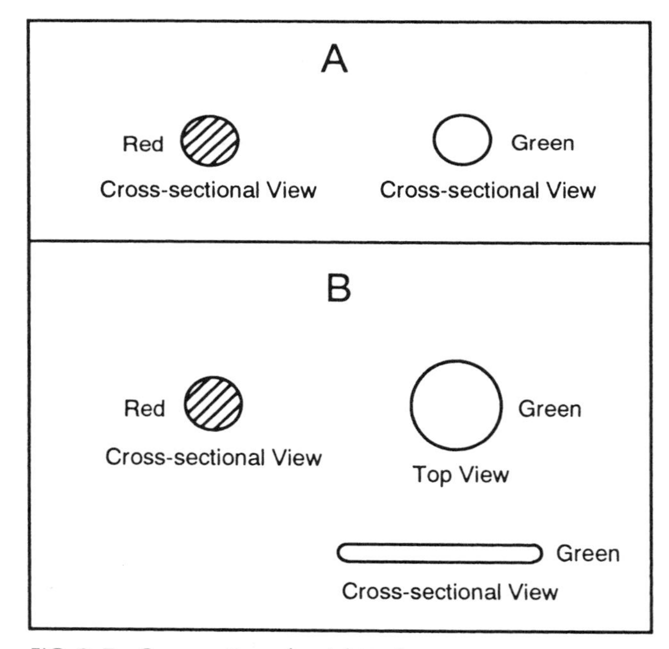

FIG. 3–7 *Conservation of weight task*

The conservation tasks you have just read are neither meaningful nor functional until you employ them with children. You are probably thinking that the tasks are so simple anyone can do them. All this proves is that you are not preoperational!

ADMINISTER THE SIX CONSERVATION TASKS TO THREE CHILDREN AND RECORD YOUR RESULTS. A FIVE-YEAR-OLD, A SEVEN-YEAR-OLD, AND A NINE-YEAR-OLD WILL PROBABLY GIVE YOU AN AGE RANGE THAT WILL ALLOW YOU TO SEE PREOPERATIONAL THOUGHT AND THOUGHT DEVELOPING INTO CONCRETE OPERATIONS. AFTER THE CHILDREN YOU EXAMINE RESPOND TO THE QUESTIONS YOU ASK, BE SURE TO GIVE THEM THE OPPORTUNITY TO EXPLAIN THEIR REASONING TO YOU.

Your results will increase in meaning if you combine them with those of your colleagues. The combination of data will also show you that age alone does not determine when a preoperational learner leaves this stage. Additionally, the data combination will show you that just because a child conserves on one task does not necessarily mean that he will conserve on others. Our experience with the tasks suggests that the first conservation usually made is of number, the second of liquid amount, and the third of solid amount.

Whatever the outcome throughout the conservation tasks, the children are being evaluated on their ability to reverse their thinking, decenter their attention, cease using transductive reasoning, and develop beyond all of the other characteristics of a preoperational thinker. Children who do not reverse their thinking do poorly in science experiments; they can only observe and report what they saw. All the conservation tasks indicate a child's ability to carry out reversals of thinking; those reversals get more difficult as the complexity of the tasks increases.

EXPLAIN THE MENTAL REVERSALS IN EACH CONSERVATION TASK, AND HOW DIFFICULT AN EXPERIMENT EACH SUCCESSFUL CONSERVATION ALLOWS A CHILD TO PERFORM.

The conservation concept is a potent tool in the hands of a teacher who knows how to use it. That teacher can identify preoperational and concrete operational learners and immediately know something about their thinking processes. The type of curricula that can be used with this type of thinker can then be identified. There is no use, for example, in asking a preoperational thinker to become involved in an educational activity that requires thinking reversals; preoperational thinkers *cannot* do them. They can observe, perceive, and report their perceptions. They need educational experiences that use the semiotic functions.

Young children *cannot* mentally represent what they can do at the level of action. That means, for example, that young children can look at the roots of a plant growing into dirt, but they cannot mentally represent that those roots came from the seed. Furthermore, if asked to draw the plant and roots as they are growing, they will not be very successful. Young children can count three blocks and two blocks and when the blocks are pushed together they will probably count five blocks. But when that action is listed on the chalkboard as $3 + 2 = 5$, they cannot mentally represent that *operation* as the same as the *action* of pushing the blocks together and counting five. Perhaps the best summary statement that can be made about preoperational children is that they cannot represent cognitively what they can do in action.

You have probably been given more information about preoperational children than you wanted! There is a reason, however, for treating the preoperational period of intellectual development this thoroughly: understanding preoperational thought greatly facilitates developing an understanding of the third level in the Piagetian model.

The Third Level

Somewhere between the ages of six and seven, when they are asked the questions in the conservation tasks, most children are absolutely certain that the quantities are conserved. In fact, they look surprised that such questions would be asked. Children who can use conservation reasoning do so because they can use the mental operations of thought reversal, decentering, and seeing states in transformation, and they *begin* to use deductive and inductive reasoning as opposed to transductive reasoning. In addition to these mental operations, there are other operations in which these children can engage, such as *seriation*, *classification*, and *correspondence*. Consider teaching a child a measuring system. The conservation of length is a necessity and so is correspondence; placing a measuring stick down once corresponds to measuring that much length. To *understand* number, children must be able to seriate; they must realize that placing six after five and before seven means that six is *more* than five and *less* than seven. Classification ability must be present before children can understand that song birds, ducks, and geese are different although they are all birds. There is little doubt that the ability to use conservation reasoning requires using such operations as serial ordering, classification, and correspondence.

The foregoing paragraph can be interpreted to say that there is an entire set of mental operations that *begin* to become available to children at around seven years of age. When children reasoning at this intellectual level apply the mental operations just discussed to a problem or situation, they do so linearly. That is, they use the operations in a step-by-step fashion. They do not apply the operation to the entire situation about which they are reasoning or any part of the problem in relation to the entire problem. They reason only one step at a time. Piaget describes the operations and the manner in which they are used as "'concrete' because they relate directly to objects and not yet to verbally stated hypotheses" (Piaget, 1969). The name of this stage of intellectual development is *concrete operational*. Other terms that have been used to represent this level of Piagetian stages include *empirical-inductive* and *intuitive*. We will use the conventional Piagetian terminology for our discussions.

Piaget's pointed remark that these operations are concrete because they relate directly to objects is a directive to schools about what they should ask children to do. Students cannot learn unless they deal with reality. Actual objects, actual events, actual situations are mandatory. In many elementary school science programs, children are asked to "learn" about the atom. Now, to understand the atom and what it means, you must understand data that support the existence of the atom. No one has ever seen an atom; it is abstract, not concrete. Concrete operational learners cannot understand abstractions. They do understand data from reality. Concepts that are built from concrete experiences are *concrete concepts*.

All school subject matter must be evaluated within the frame of reference of concrete and abstract. If the material the children are to learn can be derived from reality—actual objects, events, or situations, or all three—children in the concrete operational stage can understand it. If the material is not from reality, the children interacting with it will not experience learning. *Do not believe* that children will learn concepts if they are given only the data from some concrete event, object, or situation, or all three. They must become involved in gathering the data, evaluating the data they have been given, discussing the sources that produced the data, or in some way have an actual concrete experience with the particular data that are to lead them to the concept to be learned. You should recognize this as the learning cycle introduced in Chapter 1. The most beneficial experience is, of course, gathering the data themselves; vicarious experiences do not have the efficacy that actual experiences have. We believe there is much nonconcrete content in many elementary school science programs that needs to be eliminated.

Piaget originally stated that children begin to develop the concrete operational stage between six and seven years of age. The original investigation carried out by Piaget and his colleagues led them to state that children begin to move out of the concrete stages of thought "at about 11 or 12 years" (Piaget, 1963). In 1972, however, Piaget wrote an article in which he stated that the students used in the original investigation were taken from "the better schools in Geneva" (Piaget, 1972). He further stated that he felt the ages of eleven to twelve years, which he had concluded were the ages when children began to develop the post-concrete thought period, were derived from data taken from "a somewhat privileged population." He proposed, rather, that the development of postconcrete thought begins "between 15 and 20 years and *not* 11 and 15 years."

Our own research data corroborate Piaget's statement that concrete thought is prevalent in many students even older than fifteen years (Marek, 1992). We find consistently that over 80 percent of seventh- and eighth-graders are concrete operational, three-fourths of tenth-graders are concrete, and over one-third of the college students in our studies are concrete operational. Such data vividly demonstrate that teachers must thoroughly understand concrete thought. It usually begins in the late first or early second grade and persists in the majority of students through high school.

The presence of concrete reasoning means that actual experience with those concepts that are to be learned is the *only way meaningful understanding develops*.

EARLIER IN THIS CHAPTER WE MADE THE STATEMENT THAT YOU WOULD UNDERSTAND WHY SO MUCH TIME WAS BEING SPENT

ON THE PREOPERATIONAL STAGE OF THINKING. STOP READING AT THIS POINT, PICK UP YOUR PENCIL, AND ORGANIZE YOUR THOUGHTS ABOUT WHY UNDERSTANDING THE PREOPERATIONAL STAGE IS SO IMPORTANT. RECORD THOSE THOUGHTS.

The Fourth Level

Preoperational thinkers often indulge in the wildest kind of fantasy, which often has no basis in fact. If the world does not suit them, they just imagine the type of fanciful world they want. Concrete operational thinkers, however, are concerned with the actual data they extract from objects, organizing these data, and doing mental operations with them. These learners do not formulate abstract hypotheses from their experiences; they confine their thinking to events in the real world. They can classify, compare, seriate, and perform all the various thinking acts that will lead to the extraction of information from objects if they are given experience with concrete objects. In short, individuals in the concrete operational stage of thinking rarely depart from reality, as do preoperational thinkers, even though these departures have no lawful or logical basis.

At some time between fifteen and twenty years of age, many people find they can do a type of thinking that is not completely dependent upon reality, but which depends instead upon "simple assumptions which have no necessary relation to reality or . . . beliefs" (Piaget, 1963). According to Piaget, such an individual "thinks beyond the present and forms theories about everything, delighting especially in considerations of that which is not" (Piaget, 1963).

These quotations describe a kind of thinking that has most often been called abstract thought. Those using this kind of thinking do not need to have direct experience with reality; they can assume that it exists and use that assumption as though it were reality. Piaget states that a person performing abstract thinking has become capable of *hypothetico-deductive* thought, which consists of implications and contradictions established between propositions. Such individuals can think about the consequence (implications) of their thinking.

Another descriptive term that could be used for this stage of thought—hypothetico-deductive thinking and thinking with assumptions—is *propositional reasoning.* A proposition says the following: *If* the assumption or deduction (about such and such) is true, *then* it follows that such and such is also true; *therefore* this or that action is dictated or suggested. In other words, thought on this level in the Piaget model has a particular *form*. Here again the title given this stage is descriptive. Piaget has called it *formal operational.*

Quite evidently, formal thought is much more sophisticated than concrete thought. Inhelder and Piaget (1958) describe this difference in sophistication as follows:

> Although concrete operations consist of organized systems (classifications, serial ordering, correspondence, etc.), they proceed from one partial level to the next

> in step-by-step fashion, without relating each partial link to all others. Formal operations differ in that all of the possible combinations are considered in each case. Consequently, each partial link is grouped in relation to the whole; in other words, reasoning moves continually as a function of the structured whole.

In this quotation, the "whole" must be interpreted to mean the entire problem, event, or situation the learner is attempting to understand. A good example is the structure of the atom. The structure means nothing if one cannot see the role of each part of the atom—electrons, protons, neutrons, and so on—in relation to the entire atom. The atom is the structured whole. Concrete thinkers can see that negative charges attract positive charges, and they can also vaguely understand the role of electrical-charge repulsion. They then ask why the negative charges around the positively charged nucleus are not pulled into the nucleus. They cannot see each part in relation to the whole; they reason only one step at a time. Since the atom is a concept that requires formal thought, it is a *formal concept*. Formal thought takes into account the total number of *possible* combinations in any problem, event, or situation; a *combinatorial system* is present. This formal ability is called "reasoning with the structured whole."

Formal thought is needed to see relationships among the force, weights, and distances in a first-class lever and to visualize the distance between the earth and the other planets in the solar system. Understanding the description of the DNA molecule—often presented in books for the upper elementary grades and junior high school—also requires using the structured whole found in formal thought.

To understand an idea that has no necessary relationship to reality, a student must be able to use formal operational thought. The key word in distinguishing concrete operational thought from formal operational thought is *reality*. The concrete operational thinker can think *only* about reality—experiences the thinker is having or has recently had. The formal operational thinker "is concerned with reality, but reality is only a subset within a much larger set of possibilities" (Phillips, 1975). That larger set of possibilities exists because those capable of formal operational reasoning can think on the basis of assumptions. Formal thought is capable of departing from reality, but those departures are lawful and based on assumptions. Reasoning from assumptions—whether or not they are true is unimportant—is as legitimate to formal thought as reasoning from reality is to concrete thought.

Movement Through the Levels

By now the type of science content appropriate for the elementary school should be apparent. Children begin to display concrete thought in late first or second grade. Before that time, the science experiences provided for them must lead children to observe and report what they have seen. Since preoperational children do not see states in a transformation, doing complex experiments in kindergarten or first grade is nonproductive. But some simple experiments can be done in late first grade. A group of first-graders thought that water disappeared from a small aquarium because the fish drank it. They built a second aquarium just like the first but without fish and observed it for a period of two weeks. This was a simple experiment. Concrete operational students can do experiments that become increasingly complex as grade level increases, but they *must* have direct experience with the materials, and the concept being taught must be drawn directly from the data that the experience produces.

When children have just begun to develop formal thought they will succeed with the conservation of volume task (see the Appendix). We administered the conservation of volume task to seventeen randomly selected sixth-graders from one school and sixteen randomly selected sixth-graders from a second school in the Norman, Oklahoma, system. Of that group, twenty-nine were twelve years old, and four were thirteen years old. Only six—eighteen percent—of these thirty-three students were successful with the conservation of volume task. These data show that students can *begin* to enter formal thought before fifteen years of age but that very few do.

What happens or can be made to happen that will encourage a child to begin to develop the thought processes of the next level of cognition? Piaget (1964a) lists several factors that contribute to movement from stage to stage—maturation, experience (physical and logical-mathematical), and social transmission. He contends that any one of the three factors is not enough by itself to account for sufficient change to encourage a child to develop the next level of thought.

The simplest definition of *maturation* is the process of maturing or growing. In relating growth to the increase of intelligence, what has to grow and mature is the nervous system. The maturation of the nervous system gives each of us the opportunity to see, hear, feel—in short, to experience—more and more objects, events, and situations within our environment. Maturation, then, is the physiological or organic growth of the organism—the individual.

There are two types of *experience* that influence intellectual development. The first is *physical experience*, which includes simply poking, breaking, lifting, and squeezing objects; putting objects into water; throwing objects into the air; and engaging in all manner of activities that allow learners to gather data about their physical environment. Walking in the mud is as important as hearing pleasant sounds. There is, however, a second kind of experience, which Piaget calls *logical-mathematical*. He believes that through this experience, "knowledge is not drawn from the objects but it is drawn by the actions effected upon the objects" (Piaget, 1964a). Earlier, we gave the example of the young child who discovered that, regardless of how stones were arranged, there were always ten. These experiences require mental operations and have to do with the logic and order of the environment. Discouraging children from touching, feeling, and interacting with the environment in the classroom deprives them of the needed physical experience that leads to mental development. Asking children not to use objects and to think only about numbers provides them with an impoverished logical-mathematical experience.

Transmission means to pass along. *Social transmission* means to pass along what your own society is like, and the most common kind of social transmission is oral language: talking. But social transmission also occurs through the institutions of society—schools, churches, museums. Indeed, everywhere a child can interact with society, social transmission and concomitant intellectual development are taking place. The interaction concept is so important to us that we use social interaction as a synonym for social transmission.

There is also a fourth factor that influences movement from stage to stage in the Piagetian model. This factor, known as *disequilibrium,* will be discussed later in this chapter.

PROHIBITING CHILDREN FROM TALKING AND HAVING INADEQUATE MATERIALS AFFECT A CHILD'S OPPORTUNITIES FOR MOVING THROUGH THE STAGES IN PIAGET'S MODEL OF INTELLECTUAL DEVELOPMENT. WRITE A PARAGRAPH EXPLAINING THESE EFFECTS.

School experiences can accelerate the movement of children from the preoperational to the concrete operational stage and the movement of junior high school students and college students from the concrete to the formal operational stage. Whether or not such acceleration is desirable is questionable. (Piaget has called this "the American question.") We believe that whether or not acceleration is desirable depends on what is meant by acceleration. Most assuredly, young children can be trained to give satisfactory answers to the conservation tasks. We once trained a group of kindergarten children on the conservation of liquid task. They became so proficient that regardless of the liquid or containers used, they always gave the correct responses. Next, we used salt instead of a liquid. Salt takes the shape of the container and pours just as a liquid does, yet when salt was used the children failed the task. This experiment demonstrates that merely providing correct answers on the conservation tasks does not lead to intellectual development.

Schools teach content in science, mathematics, social science, and the other disciplines. Our position is that if acceleration occurs through the study of the disciplines taught, it is a positive experience for children. But to accelerate a child for the sake of acceleration is not, in our judgment, advisable. How then can the discipline of science be used to lead students simultaneously to learn that discipline and to increase their levels of intellectual development? The data needed to answer that question will be presented later in this chapter. Before presenting those data, some ground rules need to be established.

The first of these ground rules relates to what we call learning. Not infrequently, adults—including some teachers—believe that just because learners can repeat something, they understand it. Thus, such demonstrations as counting, saying the alphabet, writing a name, reciting a poem, repeating the multiplication tables, or naming the parts of a plant are often mistaken for learning; and they may be. Such intellectual tasks as these, however, can simply be the results of repeating something often enough until it is memorized, which represents training rather than a basic understanding of the material. This assertion can easily be checked by asking a child who demonstrates the ability to recite, for example, the multiplication tables, *why* 6 x 9 = 54. If the child explains that 6 x 9 means six added nine times, or nine added six times, and that the total is 54, learning and understanding are present. If such an explanation is not forthcoming, the child has been *trained to recite, not educated to understand.*

In our judgment, learning implies meaningful understanding. Therefore, unless understanding of the type found in the above example is present, learning has not taken place. There are many tasks in our society that require only training to perform. There are also tasks in school for which training is necessary. Suppose you wish to show your fourth-grade class how to use simple microscopes. Taking the time needed to explain the lens system in the microscope is not advisable for two reasons. First, the operation of the lens system is a formal concept and cannot be understood by concrete operational fourth-graders. Second, knowing how the microscope works is not the purpose of introducing the instrument; the purpose is to lead children to understand how to place an object under a microscope to view it. The conclusion is obvious: train students in how to use the microscope. Understanding lenses *at this point* is not necessary. But remember, training *does not* necessarily lead to an understanding of what the students are being trained in, *nor is it intended to.* Likewise, we believe that teachers should be *educated* and that to discuss "teacher training" is incorrect.

The second ground rule deals with the reality of the title of this chapter. The title implies that you really can know how children learn. This is not true. We make judgments about how children learn by observing how they act when exposed to a given object, event, or situation. From our observations we abstract patterns, which lead us to postulate how learning takes place. In short, we build a *model*, or theory, of how learning comes about. This chapter deals with the explanation of such a model.

The third, and last, ground rule is related to the first one. We have *constructed* all of our understanding for ourselves. From birth to death we are *knowledge constructors*. No one can give us *our* understandings; others can give us *their* understanding but not until we learn how to function with a particular object or in a particular event or situation have we developed our *own*. Two names are often used to identify this idea: the *development learning* model and the *knowledge construction* model. Learning how to function with a particular event or situation is really learning how to construct your own knowledge about it. How does the process of knowledge construction occur?

A Learning Model

Suppose you asked a five-year-old to compare two automobiles. The child would probably compare their size, color, and the other properties obvious to a preoperational child. A concrete operational thirteen-year-old may talk to you about tire size, speed, engine size, and other properties the five-year-old did not mention. If you next asked an engineer to compare the same two automobiles, you would probably get a completely new set of comparisons.

Each of the three observed the same two automobiles: why were different properties compared? Each of the three functions at a different level intellectually and was, therefore, concerned about, or interested in, different properties. Each saw different properties because each had different mental abilities, procedures, or systems to use in processing the data received from the environment. The notion of how data from the environment are processed is central to the theory of Piaget. "Environment" is used as an umbrella term in this chapter to mean whatever the learner is involved with at a particular moment. Thus, an infant's input data might come from a rattle, while a third-grader's data might come from a frog. Both are represented here by the term "environment."

From birth each of us develops mental processes to use in dealing with incoming data from the environment. Piaget calls these mental data-processing procedures *mental structures*; the differences in mental structures distinguish one intellectual level from another. As children move through the intellectual stages in the Piagetian model, they can process more and more complex data from the environment. This explains why the child, the adolescent, and the adult would probably give different and increasingly complex comparisons when observing two automobiles.

Mental-structure building probably begins at birth, but early structures are simple. When a baby looks at a toy and picks it up, the baby becomes able to repeat that action with any object that she or he can grasp and lift. A system for looking-grasping-and-picking-up has been established. This system is a small mental structure that Piaget calls a *scheme*. The quality of the schemes that develop early is undoubtedly dependent upon the quality of the nervous system we inherited from our parents. Notice, however, that the baby's scheme included the stimulus (looking) and all the other processes that led to the complete act. As a child grows older, she constructs more and more schemes; these eventually become integrated with each other and form cognitive structures. In other words, the basic, or generic, unit of the cognitive structure is the scheme.

As more and more cognitive structures are built, more and more data from the environment can be incorporated into them, and the individual *moves through*—according to the Piagetian model—the intellectual stage *and/or* into the next one. The process of incorporating data into existing structures is known as *assimilation*. A teacher must keep in mind that only learners themselves can assimilate incoming data from the environment; *no one can do it for them*. The learner must experience the hardness of a brick, watch mealworms go through a life cycle, or observe a lens form an image. Then and only then does assimilation take place. *Giving a child information does not lead to assimilation.*

Suppose a preoperational child is asked to assimilate a concrete operational idea or concept. The schemes of the child will not permit that assimilation to occur. But some assimilation probably does occur. Preoperational learners construct preoperational understandings of those parts of the idea that they can observe, and the child's schemes change—transform and modify—the input from the environment. Most certainly, this is not the type of understanding that the teacher, or other adults, intended should be achieved. Such an occurrence is generally responsible for adult exasperation with young children when they do or say things that are in direct contrast to what the adult believed they understood. The child heard the language, but did not assimilate the idea. This does not upset the child; rather the child is probably concerned that something is wrong with the adult but does not, cannot, know what is wrong. That is, the content of the event does not concern the child, but the social outcome in relation to the adult may.

The concrete operational child, however, is concerned about the content of an event because now this individual is probably aware that what has been assimilated is not being understood. No such awareness exists with preoperational learners. They believe that what they assimilated from the object, event, or situation was what they should have assimilated. But concrete operational learners are aware that there is a mismatch between their mental structures and what they have assimilated, and their awareness causes them concern. These learners are in what Piaget called a state of *disequilibrium*, which is revealed by such statements as "What's that?" "How did that happen?" "Why did that object turn brown?" These kinds of statements indicate that the learner is concerned about what is under consideration—an object, event, or situation—and does not fully understand it. Do not equate disequilibrium with frustration; they are not synonymous. But if disequilibrium is allowed to persist, frustration can develop, and the learner's interest will probably disappear.

Just as mental structures or schemes can change input from the environment, input can also cause mental structures to change. The disequilibrium caused by the mismatch of input and mental structures can cause new schemes to be built, or structures to be modified, combined, or both, to enable an altered structure to emerge. The entire process, the adjustment or change of mental structures, is called *accommodation* and is brought about by the disequilibrium that resulted from assimilation. But when structures have been adjusted to accommodate the new inputs (or intrusions) from the environment, the learner has once again reached a stage of equilibrium.

DESCRIBE A SITUATION IN WHICH YOU MADE AN ASSIMILATION AND FOUND YOURSELF IN DISEQUILIBRIUM. WHAT DID YOU DO TO PUT YOURSELF BACK IN EQUILIBRIUM?

The procedure by which mental structures are revised is the process of equilibration. Revised mental structures make possible assimilations that were not possible before, and represent the movement of the learner more deeply into a particular intellectual stage or into an advanced stage. The child can now undertake learning that could not have been accomplished before the new, revised structures were present.

The new learning, however, depends upon accommodation, which is the result of disequilibrium; but disequilibrium is caused by assimilation. In the truest sense, therefore, without assimilation there is no accommodation and no new learning. Remember, however, that the *impetus promoting* accommodation is *disequilibrium*.

Teachers need to be aware that the key to the entire process of bringing about disequilibrium is assimilation, and that this must be done by the learner—no one else can do it for the individual. To emphasize that disequilibrium is focused on the learner, Piaget (1969) has frequently used *self-regulation* as a descriptor of equilibrium. By using the theory base of Piaget to develop a view of teacher responsibility in promoting learning, it is clear that the teacher's first responsibility is to involve the learner in an assimilation that will cause disequilibrium.

Assimilation and accommodation cause a change in the learner's mental structures, representing an *adaptation* of the learner to the inputs received from the environment. But adaptation is accomplished only when there is a balance or equilibrium between accommodation and assimilation. According to Piaget (1963), the learner has found an "accord of thought with things." That a learner makes an adaptation and puts thought in accord with things—inputs from the environment—is certainly at least a part of learning.

But no thought exists in our cognitive structures by itself. Every adaptation that results from assimilation and accommodation is always related to all other earlier adaptations that resulted in mental structures. In other words, the new structure is placed among all the other structures in some type of mental-structure *organization*. An obvious definition for organization is the relationships that exist between a new mental structure and previous mental structures. Piaget (1963) defines organization as the "accord of thought with itself."

EXPLAIN WHY ORGANIZATION AND ADAPTATION ARE COMPLEMENTARY PROCESSES.

Earlier, we made the point that the impetus promoting mental-structure construction and reconstruction is disequilibrium. Disequilibrium is *caused* by assimilation and *causes* accommodation. In Piaget's theory, equilibrium is dynamic, not static. It is consistently being caused or causing. In addition, as disequilibrium diminishes, the newly constructed or reconstructed structures become more and more stable, and we begin to see inconsistencies and gaps in them that we did not see before. These gaps and inconsistencies are perhaps what lead us to say "yes, but . . ." after we believe we have mastered a new concept. The "yes, but . . ." response can cause a new assimilation, which in turn can produce a new disequilibrium. Phillips explains the entire process in this way:

> Each equilibrium state . . . carries with it the seeds of its own destruction, for the child's activities are thenceforth directed toward reducing those inconsistencies and closing the gaps. (Phillips, 1975)

Piaget (1963) refers to assimilation, accommodation, adaptation, and organization as the functional *invariants* of intelligence. By that he means that, regardless of the age of the learner, the process is the same. It begins with assimilation and continues through organization. The material with which the human organism functions and the sophistication with which functioning occurs change as a learner gets older; these, of course, are dependent upon the learner's intellectual stage. The invariant relationships that exist among assimilation, accommodation, and organization are shown in Figure 3–8. Those relationships are called the intellectual *function*. Although we cannot establish that Piaget ever referred to functioning as the learning process, we look upon it as such.

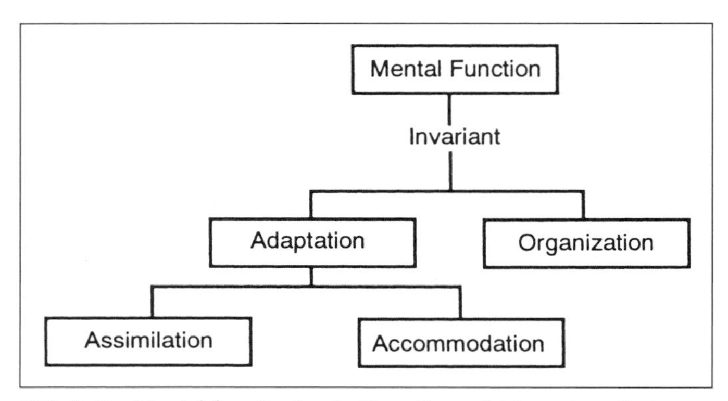

FIG. 3–8 *Mental functioning in Piaget's model is an invariant.*

While the process of functioning itself is an invariant, mental structures are not; they are consistently changing. Those changes probably begin the instant we are born, continue throughout our entire lives, and are responsible for our movement through the intellectual stages. To us, mental structures represent learning. Now, when we ask children to "learn something" we have asked them to transform the inputs from the environment into their own mental structures. This will happen only if each child has a system that can make the required transformations. Suppose children are to be taught how to weigh a liquid. They need to understand that the beaker must be weighed empty, then with the liquid in it, and the two values subtracted. Unless children can use the necessary mental operations to make all of the transformations required in this process, they cannot understand that at the end, they will know the weight of the liquid. Using mental structures to make such transformations on the data received from the environment represents to us the process of learning; knowing how to describe the mathematics involved in a first-class lever does not. A teacher must understand the type of mental structures available to learners at different ages.

Think back to the characteristics of the preoperational child. The mental schemes the preoperational child is able to use are the semiotic functions. These schemes allow learners to observe and report, but what they report they observe is not necessarily what a concrete or formal operational observer would report. The mental structures available to the concrete operational learner are represented by serial ordering, conservation, correspondence, and classification, used in a linear, step-by-step fashion. The formal operational thinker functions with the structured whole. Mental structures, therefore, are mental operations used in transforming into meaning the data received from our environment. According to Piaget, "Operational structures are what seems to me to constitute the basis of knowledge. . . . The central problem of intellectual development is to understand the formation, elaboration, organization and functioning of these structures" (1964a). It is this "central problem" that we classroom teachers must learn to solve, but we must also learn to implement a solution. In other words, how do we base classroom teaching upon the Piagetian functioning model?

WRITE WORKING DEFINITIONS OF ASSIMILATION, ACCOMMODATION, ADAPTATION, DISEQUILIBRIUM, ORGANIZATION, AND FUNCTIONING. KEEP YOUR DEFINITIONS IN A CONVENIENT REFERENCE PLACE WHILE YOU STUDY THE REMAINDER OF THIS SECTION.

Up to this point, we have examined two factors concerned with learning. The first is mental functioning, which is invariant. Intellectual functioning led to the second factor concerned with learning: the formation and reconstruction of schemes and mental structures. But there is also a third factor related to mental structures and functioning that we have not considered.

This factor can best be understood by considering data that resulted from administering the conservation of volume task (see Appendix B) to 1,108 students in grades ten through twelve. Just as the conservation of volume task directs, two identical glass cylinders were filled to the same height with water. Each of the students was given two solid metal cylinders having exactly the same shape and size but different weight. The students were told that the two metal cylinders were going to be put into the partially filled glass cylinders, that the metal cylinders would sink, and that this would result in the water level rising. They were then asked, *before* the cylinders were submerged, whether one of the metal cylinders would push the water level up more and which cylinder would do this, or if the metal cylinders would push the water levels up equally. A total of 484 of the 1,108 students (44 percent) stated that one metal cylinder would push the water level up more than the other.

This example demonstrates what Piaget referred to as *content*. In the conservation of solid amount task using clay balls, for example, young children often state that there is more clay in the pancake. The results of these two conservation tasks tell the interviewer that the children have assimilated the situation and how the interviewer's *content* directed them to behave. Do not be misled into believing that the proper questions were not asked or that the two subjects did not understand what they were expected to do. Neither redirecting the question nor making the questions "smaller" will change the responses, which were dictated by content. To change the response, the entire functioning, structure-building process must be activated and disequilibrium brought about. In other words, the content factor is a variant inextricable from the structure and function portions. We have found it extremely useful to *think about content as the way the child believes the world works*. A young child really believes that flattening a ball of clay produces more clay, just as an adolescent believes that a heavy object pushes down harder on water than a lighter object, and moves the water level up more. Those responses represent content, the way the individual believes the world looks.

Mental structures, content, and function, Piaget postulates, represent *intelligence*. According to Piaget, intelligence is not a *static* attribute that each of us possesses. Rather it is a *dynamic* factor, which changes through the construction and reconstruction of mental structures and content that begins with assimilation, disequilibrium, and accommodation. We have diagrammed the Piagetian intelligence model in Figure 3–9. The diagram was not prepared by Piaget or his associates. It is our interpretation of the relationships among the factors that constitute intelligence; that is, mental structures, content, and functioning. Reading the diagram from top to bottom explains the overall intelligence theory, but when implemented through the learning cycle in a classroom, the process begins with assimilation.

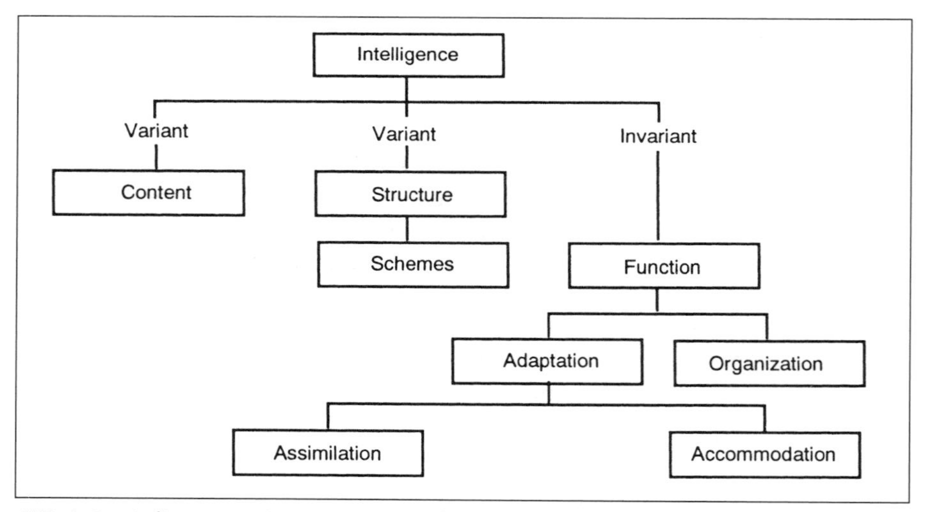

FIG. 3–9 *A diagrammatic interpretation of Piaget's model of intelligence*

PREPARE EXPLANATIONS OF CONTENT, STRUCTURE, AND FUNCTIONING THAT ARE SATISFACTORY TO YOU. NOW COMPARE YOUR EXPLANATIONS WITH THOSE OF YOUR PEERS. AGREE UPON SUITABLE EXPLANATIONS OF THOSE CONCEPTS.

Combining Models

Earlier in this chapter, we considered Piaget's model of intellectual development, often called the *stages* or *developmental model*. Then we explored the intelligence model. Certainly these two factors—level of intellectual development and intelligence—are related. We have found that performance on the Piagetian tasks to measure formal operational thought (see Appendix B) is positively correlated with a measurement of static IQ; the correlation coefficient is 0.44. But that correlation does not relate to the relationship between the intellectual development and intelligence models discussed in this book.

To uncover this relationship, you must think about what the stages of intellectual development really mean. While you are doing that, however, remember that mental functioning is an invariant—we always assimilate, disequilibrate, accommodate, and organize. How does assimilation by a preoperational child differ from assimilation by concrete and formal operational thinkers, who assimilate different aspects of the same object, event, or situation? Why are their assimilations different? They have different schemes and mental structures to use in assimilation; that is, they have different systems to use in transforming data received from the environment.

Furthermore, each person thinks the world works in a different way. The preoperational child believes that pouring water from one size of container to another increases or decreases the amount of water. The concrete operational child thinks that such a suggestion is silly. Of course, the amounts are the same: "All you did was

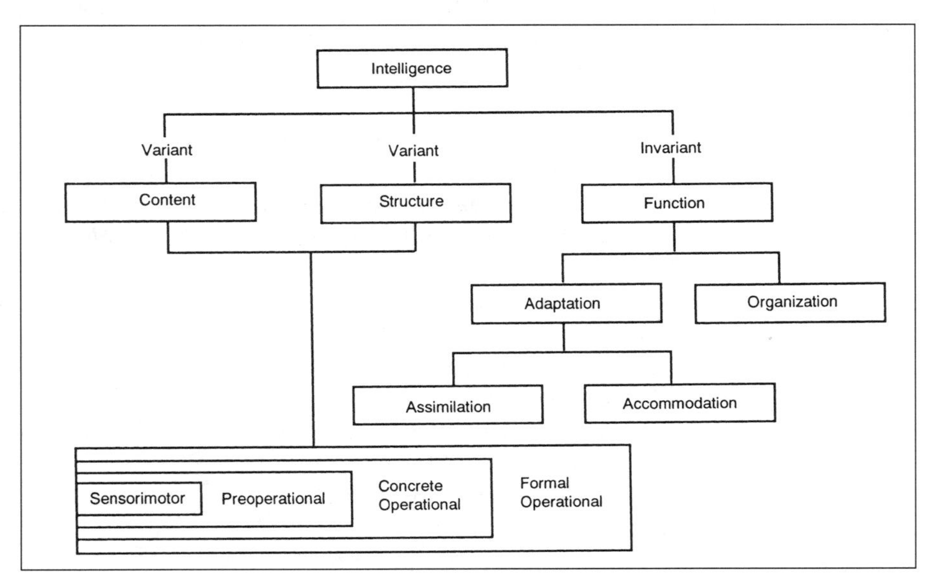

FIG. 3–10 *An interpretation of the relationships within Piaget's model of intelligence*

pour it from one container to another, you did not add any or take any away." That same concrete operational child, however, will also tell you that heavier objects "push down" on water more than less heavy objects do and push the level of the water up more than less heavy objects, even when all other variables—size, shape, height of water—are the same. The formal operational thinker considers this latter task rather silly because "It's the volume, not the weight, that controls how much water an object displaces." In other words, the content is different for each of the three.

Content and mental structures determine intellectual level. These mental structures, however, also control what can be assimilated in order to change content and mental structures. Perhaps considering this will make it easier to see why content and mental structures are called variants. Our interpretation of the relationship between the intellectual development and intelligence models is shown in Figure 3–10 on page 65.

References

FLAVELL, J. H. 1963. *The Developmental Psychology of Jean Piaget*. Princeton, NJ: Van Nostrand.

INHELDER, B., AND J. PIAGET. 1958. *The Growth of Logical Thinking*. New York: Basic Books.

MAREK, E. A. 1992. "Conceptualizing in Science: Misconception Research Using a Constructivist Model." In *Literacy Across the Curriculum*, edited by C. Hedley, D. Feldman, and P. Antonacci, 266–285. Norwood, NJ: Ablex.

PHILLIPS, J. L., Jr. 1975. *The Origin of Intellect: Piaget's Theory*. 2d ed. San Francisco: Freeman.

PIAGET, J. 1951. *Play, Dreams and Initiation in Children*. New York: Norton. Original French edition published in 1945.

———. 1963. *Psychology of Intelligence*. Paterson, NJ: Littlefield and Adams.

———. 1964a. "Development and Learning." *Journal of Research in Science Teaching* 2(3): 176–186.

———. 1964b. *Judgment and Reasoning in the Child*. Paterson, NJ: Littlefield and Adams.

———. 1969. Foreword to *Piaget and Knowledge: Theoretical Foundations*, by H. G. Furth. Englewood Cliffs, NJ: Prentice Hall.

———. 1972. "Intellectual Evolution from Adolescence to Adulthood." *Human Development* 15(1): 1–12.

PIAGET, J., AND B. INHELDER. 1969. *The Psychology of the Child*. New York: Basic Books.

PART 2

Term Introduction

4 The Theory Base of Elementary School Science

Part I of our text, *Exploration*, was constructed to guide the reader in gathering data for these questions:

1. What is the nature of science and how is science to be taught?
2. What are the goals of science education?
3. What is the nature of the learner?

Part II, *Term Introduction*, will lead the reader toward developing an understanding of the relationships or linkages among the nature of science, the goals of science education, the nature of the learner, and the learning cycle. In other words, this chapter is designed to guide the reader in constructing a *theory base* for elementary school science teaching.

A Teaching Model

In Chapter 3 you explored a learning model called the *developmental* or *knowledge construction* model. We believe that a teacher cannot teach students unless that teacher understands a model of how students learn. Without that understanding, all a teacher can do is to expose students to content and, using tricks and "motivational techniques," hope they will glean enough from the exposure to pass the tests. This can happen when a teacher does not have a theory base about learning upon which to base teaching procedures that will promote learning. The learning model you have just explored is such a theory base. Where does a teacher start planning to use the theory base?

The knowledge construction theory base begins with the process of assimilation. During assimilation, learners acquire all the information they can about what is to be learned. We believe that in science, or any discipline, students learn concepts. Remember from Chapter 1, concepts are the major ideas in a discipline, such as "Energy makes things happen in our environment," "Chemical change cannot be reversed," "Food producers, food consumers, and decomposers constitute a community." During assimilation, students absorb the essence of concepts such as these. So the teacher must first ask, how do you promote assimilation?

If preoperational and concrete operational learners are to absorb the essence of a concept, they *must* experience that concept. Students must use materials that they can touch, feel, hear, and observe; they must use every sense possible in finding out everything they can. The materials relating to the concept must be made to interact with each other; frequently the teacher needs to supply directions about how this should be done. These directions do *not* tell the students what they should learn from the materials, however, nor does the teacher *explain* the concept to be learned at the beginning of the assimilation. Students are simply exploring the materials on their own or by following teacher-provided directions. This is the exploration phase of the learning cycle and we believe that exploration of the materials promotes assimilation.

If students are assimilating, then disequilibrium can occur, and the exploration phase fosters assimilation and disequilibrium. The next phase, term introduction, provides for accommodation or the construction of new mental structures. These new mental structures allow for the development and understanding of the science concept(s) inherent in the operant learning cycle—the student is reequilibrated. The concept-application phase of the learning cycle is designed to allow students to relate their newly learned concept to other concepts or to apply this new concept in other situations. Concept application corresponds to the process of organization in Piaget's model of mental functioning. The learning cycle's derivation from Piaget's model of mental functioning is illustrated in Figure 4–1.

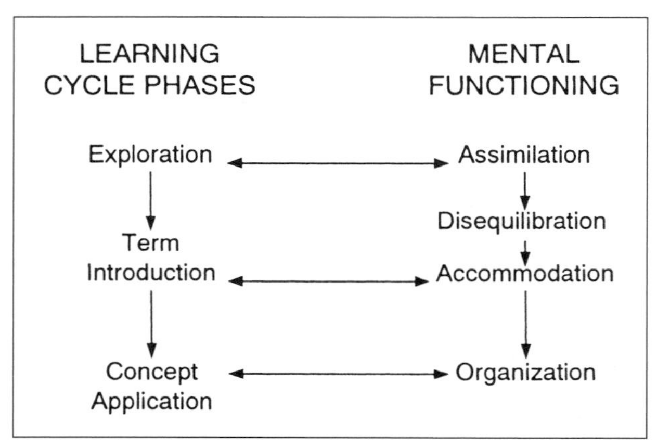

FIG. 4–1 *The learning cycle and Piaget's model of mental functioning*

Development and Learning

In the following passage, Piaget (1964) clearly explained the relationship between development and learning:

> So I think that development explains learning, and this opinion is contrary to the widely held opinion that development is a sum of discrete learning experiences. For some psychologists development is reduced to a series of specific learned items, and development is thus the sum, the cumulation of this series of specific items. . . . In reality, development is the essential process and each element of learning occurs as a function of total development, rather than being an element which explains development.

Development explains learning; what a potent educational idea! Preoperational children can learn only content and *school stuff* that uses preoperational thought, not concrete or formal operational thought. Concrete operational students *cannot* comprehend content that demands formal operational thought. Why is this true? The answer lies in two elements of the Piagetian intelligence model presented in Chapter 3, *assimilation* and *mental structures.*

Think about the meaning of assimilation developed in Chapter 3. Assimilation is the process by which learners—school or nonschool—take in or perceive what they are experiencing. The learner becomes aware of what is expected during the process of assimilation. Assimilation was accomplished by the mental structures described in Chapter 3 as systems of transformation. Now, transformation to us means changing, adjusting, or in some way mentally manipulating what has been assimilated so that it has meaning. During the entire process, the learner is in a state of disequilibrium that causes mental structures to change: accommodation takes place and equilibrium is restored. However, if the learner's mental structures—or schemes—do not have the quality that permits transformation, no disequilibrium and accommodation occur, or disequilibrium occurs and the student can do nothing about it. In this instance, teachers frequently provide the content in a form that can be memorized for recitation or repeated in an examination; but no understanding is present in students' mental structures because no accommodation has taken place. What has been memorized is quickly forgotten.

Preoperational children have a beautiful trait that prevents them from becoming disequilibrated. That trait is *transductive reasoning.* If you are unsure what transductive reasoning is, return to Chapter 3 and read about it. When preoperational children assimilate something that cannot be transformed, they seek a specific reason to explain what they cannot mentally transform. Thus, in the conservation of number task, the stack of checkers made from the row cannot be related to the row; that action cannot be transformed. The preoperational child centers attention on the height of the stack, and, by using transductive reasoning, explains that there are more in the stack because "the stack is taller." Other transductive responses preoperational children give are, "There are more checkers in the row because it is longer," and "There are more checkers in the row because it is red." The incoming data cannot be transformed; no disequilibrium occurs because of transductive reasoning.

The presence of transductive reasoning in preoperational children *probably* permits them to remain sane. Just imagine being in a world constantly bombarding you with objects, events, and situations that you cannot explain, although those around you keep asking you to. Such an environment would probably lead you to begin doubting yourself and your relationships with the world. But preoperational children do not seem to have this problem. They transform and modify the data they receive with transduction and explain away all discrepancies the world forces on them. Transductive reasoning keeps them from becoming disequilibrated. But in Chapter 3 we made the point that accommodation is brought about by disequilibrium, and that the visible or measurable products of accommodation are usually called learning. Furthermore, to move deeply into one intellectual stage and to construct the mental structures necessary to eventually develop the next stage, we have said that disequilibrium and accommodation are necessary. How, then, do preoperational children become concrete operational if they do not become disequilibrated and accommodate, as the foregoing seems to suggest?

As preoperational children interact with their environment they begin to organize what they do—their actions—in each situation. Eventually, they generalize these actions and transfer them from one situation to another. For example, when children are first introduced to a tricycle, they learn that there are certain actions they must complete if they are to make it move. They will probably meet the next tricycle they encounter cautiously but will attempt those actions they found successful on the first tricycle. Eventually, they will generalize those actions to any vehicle analogous to a tricycle. Recall that such generalizations Piaget called *schemes*, which he defined as "organization of actions as they are transferred or generalized by repetition in similar or analogous circumstances" (Piaget and Inhelder, 1969). Notice that schemes have been built from actions; the preoperational child's schemes direct actions to be done. If, however, the child were asked to describe how to ride a tricycle, that explanation would not be forthcoming because it requires mental operations involving, at least, several reversals of thought and the ability to see states in a transformation. The presence of mental operations indicates that mental structures are present; preoperational children have only mental schemes available to them.

A young boy was using a rather large syringe to blow up a balloon. He would pull the plunger on the syringe out, put a balloon on the syringe's nozzle, push in the plunger, pinch the top of the balloon to keep the air from escaping, remove the syringe from the balloon, withdraw the plunger, and start the process over. When the teacher asked him to explain how he did the process, he replied, "I don't know." He had developed the needed schemes by organizing his actions but did not have the mental operations either to mentally build the model that the experience had provided or to explain that model. Thus, the preoperational child adjusts his schemes as he repeats experiences or has an experience that closely resembles an earlier one. Eventually, the child begins to ask questions about the relationships between experiences; this child has now begun to perform mental operations and has begun to move into the concrete operational stage of development.

In summary, for anyone to learn, the content to be learned must match the intellectual level of the learner in order to enable the learner to build schemes—in the case of preoperational children—or to cause disequilibrium in concrete and formal operational students. But specifically, what do students learn from the content?

The Products of Learning

Children, experiencing a learning cycle about plant growth, planted seeds in several different containers and placed some of the containers in the light and some in the dark. They gave each container the same amount of water. The number of seeds in the dark that produced plants was approximately the same as the number of those in the light that produced plants, but after a few weeks the plants in the dark died. What would children learn from this investigation?

Children learn how to conduct an investigation. They learn how to collect, process, and analyze data and how to make generalizations from their analyses. In addition, they can frequently state hypotheses after they have analyzed the data. In Chapter 3, we said that the stating of hypotheses is a mark of formal operational thought. Concrete operational thinkers, however, can state a *type* of hypothesis that deals with what they believe will happen, based upon the experiences they have had. The data they use in making this type of hypothesis must come directly from those experiences and cannot be drawn from any abstract or hypothetical propositions. In the plant-growing experiment, for example, children will state what they believe will happen—their hypothesis—in terms of data they have collected. After stating their experience-based hypotheses, concrete operational thinkers can test those hypotheses and construct understandings about what happened in the investigation. Such understandings represent the *content concepts* that the children carry away from the investigation with them. (The young investigators also learn about the nature of science from an experience like the plant-growth investigation described here). An elementary school science investigation, therefore, leads students to understand content by leading them to construct concepts.

The Learning Cycle and the Ability to Think

In the discussion that follows, we will equate the "ability to think" with students' development and use of the rational powers: classifying, comparing, evaluating, analyzing, synthesizing, imagining, inferring, deducing, recalling, and generalizing. We first introduced this idea in Chapter 2. For another presentation of this idea, refer to "An Educational Theory Base for Science Teaching" (Renner and Marek, 1990).

Consider the exploration phase of the learning cycle. It is, of course, the time during which the major assimilation that leads to conceptual understanding takes place. In making this assimilation students classify the results they receive, which means that they compare them, and comparing results requires at least a minor evaluation. Students use several of the rational powers, therefore, in just the act of exploring. Before term introduction, students must make a thorough analysis of the data resulting from their exploration. Term introduction is obviously a synthesis incorporating the use of imagination. Classifying, comparing, evaluating, and inferring are necessary in formulating the concept. All these activities lead to transference of the data received through the context of exploration to the context of knowledge construction. Such activities also make evident why accommodation takes place during the term introduction phase.

In the concept-application phase, the transfer of knowledge from one context to another reaches its zenith. The newly acquired knowledge—the new concept—is immediately put to use in a new context and with new materials. This causes students to reorganize their fresh understanding of the concept and, of course, generalize about it. Most certainly, students are using deduction throughout this entire learning cycle phase.

The evidence we have presented here clearly leads to the inference that providing experiences with all the phases of the learning cycle leads students to develop their rational powers, making it an appropriate teaching procedure to use in the classroom. Before this teaching procedure can be used, the content the students are to study must be organized into learning cycles, as we will describe in Chapter 5. Thus, the combination of curriculum organization and classroom teaching procedures using the learning cycle leads students to achieve the *central* purpose of education, that is, they are developing the "ability to think."

WRITE A PARAGRAPH—MORE IF YOU NEED IT—THAT DESCRIBES WHY THE LEARNING CYCLE IS AN APPROPRIATE TEACHING PROCEDURE TO USE IN LEADING STUDENTS TO DEVELOP THE ABILITY TO THINK.

Science and the Learning Cycle

In Chapter 1 we developed several ideas relating the learning cycle to the discipline of science. We will expand upon those ideas as we construct the remaining portion of our model for the theory base for elementary school science.

Most assuredly, the exploration and term introduction phases of the learning cycle represent a quest. But what about concept application; are the activities in that phase representative of science? Perhaps, perhaps not. To adhere to our scientists' descriptions of science (Chapter 1), which we do, the concept-application phase of the learning cycle may not be necessary. It depends upon the nature of the concept application. The exploration phase provides experiences and the term introduction phase leads students to build a logical system.

Why, then, is the concept-application phase included in the learning cycle? There are two reasons. First, the mental functioning model of Piaget demonstrates that after a new idea has been formulated (a "logical system"), thought has been put in accord with "things." This accommodation is followed by a period in which these new thoughts are organized and integrated with old thoughts. Or, as Piaget stated, thought is put in accord with thought. You will remember from Chapter 3 that this process is called organization. Human beings will take new ideas and organize them with old ideas. Learning does not stop with Einstein's new logical system (see Chapter 1); it uses old logical systems to explain the new and in the process, revises the old systems. The concept-application phase of the learning cycle, therefore, represents how *content* ideas change as more and more concepts are learned. Those changes represent true knowledge construction.

There is a second reason for including this third phase in the learning cycle. Whenever we learn new ideas, we need to apply them to other related ideas and truly expand those ideas. This is probably another way of putting thought in accord with thought. Activities provide the opportunity each of us needs when we learn something new—we need practice with the new idea. Doing additional experiments, answering questions, and reading about the new logical system or concept *after* it has been constructed give us ample opportunity to practice. Understanding of the concept is expanded through practice and applications of the concept in new situations.

Data have been presented here to lead *you* to conceptualize that when using the learning cycle as a teaching procedure, science is being taught. The data came from several scientists and a science historian. They clearly demonstrate that students must interact with the materials of the discipline to collect data and make order out of the data. The order students produce either is a conceptual understanding or leads to it. The learning cycle, therefore, is not a method of teaching science. The learning cycle comes from the discipline itself; it represents science. If science is to be taught in a manner that leads students to construct knowledge, they must make a quest. The learning cycle leads students on that quest for knowledge.

A Persistent Complaint

Whenever science teaching is directed at implementing the "quest for knowledge," there are always complaints that less content will be "covered" than if the students read about the products science has produced in a textbook. The complaint may be accurate, but it lacks validity.

The "lack-of-coverage" complaint is invalid because letting students read about the products of science from a textbook is *not* science. What is said is being taught—science—is not taught.

There is also a second reason that the "lack-of-coverage" complaint—although probably accurate—should not overly concern teachers using the learning cycle. Earlier in this chapter we demonstrated that the learning cycle leads to the development of the rational powers. Since the rational powers are the essence of the ability to think, students who have experiences that enhance that development are in a position to carry on their own education. Fostering that ability in students is the greatest service schools can perform for them.

Since the learning cycle leads to rational power development, which will, in turn, lead students to achieve their own goals, the "lack-of-coverage" complaint seems unimportant. Content is "covered" through the learning cycle, and if you need to reassure yourself, consult Chapter 8.

WRITE A PARAGRAPH THAT EXPLAINS YOUR CONCEPT OF THE "LOGICAL SYSTEM" IDEA IN EINSTEIN'S DESCRIPTION OF SCIENCE (SEE CHAPTER 1). USE THE IDEAS JUST PRESENTED.

Activities in the Phases of the Learning Cycle

We have said that in a learning cycle students explore, experience concept understanding, and participate in applying concepts. But specifically, what kinds of *activities* do they engage in? What do students actually do while exploring, constructing, naming, and applying science concepts? What kinds of experiences do teachers provide children when teaching with the learning cycle? We believe there are six specific essential experiences that would make classroom activities recognizable to a scientist as science. These essential experiences are observing, measuring, interpreting, experimenting, model building, and predicting.

Figure 4–2 shows the relationship of these six essential experiences to the learning cycle and to science. As Figure 4–2 suggests, *Exploration, Term Introduction,* and *Concept Application* used together properly represent doing and learning science. These three major aspects can be further reduced to simpler processes, the six blocks at the base of the pyramid labeled "Essential Experiences."

The learning cycle can be depicted as two triangles and a rectangle, as shown in Figure 4–3. All six essential experiences may be used in arriving at the idea, or concept, to be constructed from the data collected during the exploration phase. The processes, although not used in equal amounts of time, are used in an order

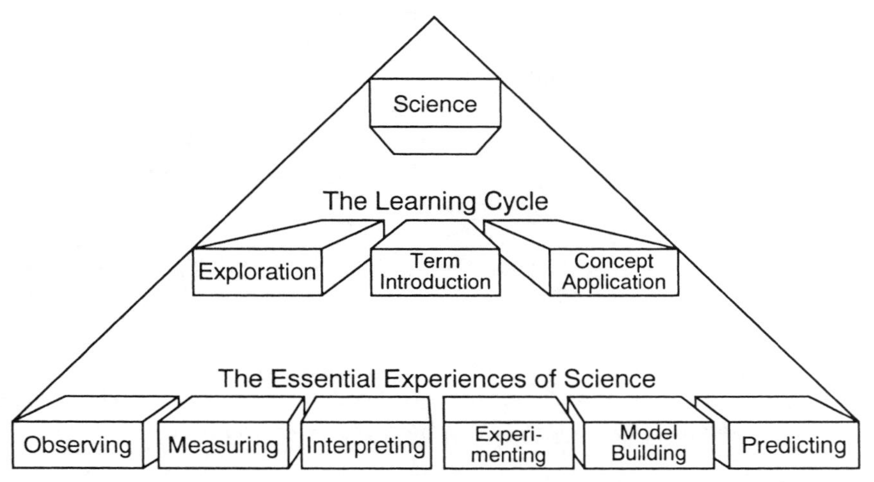

FIG. 4–2 *The basis of the pyramid of science is composed of the processes of science. These processes are essential in doing science.*

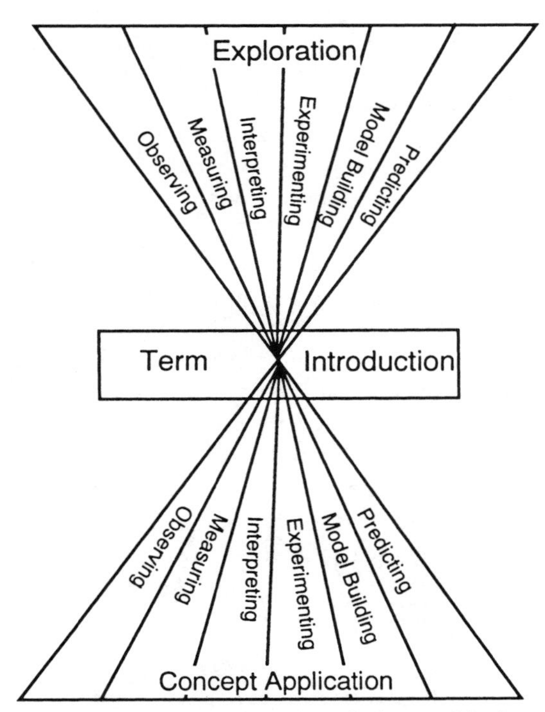

FIG. 4–3 *The processes of science are used in all phases of the learning cycle.*

or combination that will culminate in the concept. For example, when exploring a new aspect of nature, an individual would logically begin by observing and measuring to procure information and then proceed to interpreting, predicting, experimenting, and model building. It is not reasonable, for example, to begin to use model building and predicting until some information is available and some interpreting has been done. After the *concept* has been grasped, all six of the essential experiences can again be used in concept application.

Each of the six experiences is a process used in practicing science or learning science. As processes, they are tools having certain basic components that identify them, whether they are being used to gather data or apply an idea. We will explore each of the six processes by presenting first a short description and then a learning cycle that focuses on a specific essential experience. Each of the learning cycles we present provides students with more than one essential experience, but one essential experience is the primary focus of each learning cycle below.

Observing

If you are interested in acquiring information about an object you are not familiar with, the most obvious thing to do is look at it. Your observations can give you a great deal of specific information and can lead you to other investigations that will give you much more. You can make observations in many ways. Feeling, squeezing, poking, rubbing, and listening are but a few of the methods (other than visual) that can be used to make observations. Observing is the first action the learner performs in acquiring new information. The curriculum must provide opportunities for the learner to have extensive experience in making observations.

If children are going to learn how to observe, they must be given the opportunity to watch, feel, squeeze, poke, and do other things that will enable them to describe the object they are observing. Furthermore, the object they observe must be one the children feel comfortable with and not something so foreign that they are afraid of it or uncertain of what can or should be done with it. When young children go to school, they are probably more familiar with the objects in the environment than with anything else. The first observations children can be led to make, then, can be of the common things in their immediate environment. The environment contains many different shapes, but shape as a concept is usually not evident to children. The following learning cycle leads to the concept of shape. We have made the assumption that the children will not be reading, so we have written the following learning cycle for the teacher.

Exploration

The children at their desks observe wooden blocks cut into triangles, circles, squares, rectangles, and diamonds. The children learn to describe the shapes of the various wooden objects. The teacher provides the shape label as needed. Cardboard shapes may also be used. The teacher calls out a shape and each child holds up an object of that shape from the child's own tray.

Term Introduction

The teacher collects many objects that are familiar to children, such as a plate, a cup, a picture, a book, a watch, a coin, a bell, and so forth. The teacher holds up one object at a time, then asks some child to describe the shape of the object. After the teacher has gone over the collection of objects, the students or the teacher then verbally states the following idea:

Objects have a shape.

Concept Application

Have the children first find square objects in the classroom. Round, triangular, rectangular, and other shapes are found next. Have the children observe and describe objects of a particular shape on the playgrounds, along the street, and at home. Be sure that both living and nonliving objects are included.

Measuring

After taking measurements of any object, investigators (adults or children) are able to make statements that are much more definitive than those they were able to make based only on qualitative observations. We cannot, for example, look at a plant today and specifically say it has grown a definite quantity since yesterday. Our senses might tell us that the plant has grown, but they certainly would not tell us how much. In order to be able to state how much a plant has grown in three days or in a week, we must be able to refine the measurements our senses allow us to take. Our senses are not only inadequate to make a measurement as small as that of daily plant growth, they are also woefully inadequate in accurately estimating large measurements, such as the distance to the sun, the velocity of sound, or the weight of an elephant. Measurements are necessary to extend our senses down to the infinitesimal and up to something approaching the infinite, because our senses are not reliable as measuring devices except in a very approximate way.

Measurements can be considered observations, but they are quantitative observations that can be repeatedly taken in the same manner at different times. Variations will occur in measurements because of growth in a living organism (if that is what is being measured) or inaccuracies or inconsistencies that occur in the application of the measuring standard. To enable pupils to learn how to use observations from quantitative measurements, the elementary school science curriculum must provide appropriate learning experiences for pupils at all levels.

The following learning cycle has been used many times on the fourth-grade level to lead children to understand the concept of average, which must be based upon measurements. These directions are written for the students.

Exploration

Try this experiment. Mark a straight line for a starting line. Put your feet together with the toes of both feet on the starting line. In one hop, hop as far as you can with your feet together. Measure the distance you hopped in centimeters. Again, hop as far as you can and measure the distance. Make a record of the two distances.

Compare the two distances you hopped. If the distances are not the same, take a part of the longer distance and add it to the shorter distance. Make both distances the same. Compare this new distance with each of the distances you measured. How is the new distance different from each of the earlier distances?

Repeat what you did. Hop as far as you can. Then measure the distance. Hop again and measure the distance. Do this four times. After each time, make a record of the distance.

Now, take a part from each of the longer distances. Add these parts to the shorter distances until all four are the same. Make a record of this distance.

Term Introduction

You made all four distances the same. To do this, you took parts from the longer distances. You then added these parts to the shorter distances. The new distance you obtained is called the

average.

The new distance is the *average distance* you can hop. In many experiments, you will report the average result. An average often provides accurate data and is usually better than the results of only one experiment.

When you listen to people talk, you will hear the word "average" used to describe almost everything. People talk about average driving speeds, average salaries, average food prices, and even average people. There is an average number of letters on each full line of print on this page. Perhaps you can find out the average.

Averages are important in your school studies. You will be working with averages in many ways in your mathematics, science, and social science classes.

Concept Application

Try this. Add together the four distances you hopped. Divide the sum by four. The number you get is also called the average. Compare the average with the average you got by taking away and adding on.

Sometimes it is very hard to get an average by taking away and adding on. For example, suppose you wanted to know the average height of the students in your class. It would take a long time to find the average by taking away and adding on.

Think about how you could find the average height of the students. What numbers would you need to add together? What number would you divide by?

Understanding Average

1. Find the average height of the students in your class. Find the average in centimeters. Find the average in inches.
2. Find the number of students in five different classrooms in your school. What is the average number of students in those five classrooms?
3. Determine the average age in months of the students in your class.
4. Find the average age of the members of your family. Which has more variation, the ages of your family or the ages of the students in your class?

5. One boy reported that the average time he spent coming to school each morning for one week was fifteen minutes. On one of the mornings it took twenty-two minutes. Explain how he could have an average time of fifteen minutes.

Interpreting

As data accumulate, a person who wishes to use them tries to make them understandable. In an elementary school classroom, data interpretation has a variety of forms. In the learning cycle, data interpretation usually follows observing and measuring. Interpreting can be as simple as grouping objects into heavy and light or deciding if an object is rough or smooth, or as complex as building a model to explain patterns in data. Interpreting is "making sense out of data."

In almost every case, interpretation leads to experimenting and predicting. If during the interpretation of data a pattern or trend is discerned or suspected, this pattern or trend can be stated as a possible generalization or a hypothesis. A carefully planned series of observations, measurements, and experiments can then be conducted to learn more about the relationships found.

Interpreting involves the discernment of the various factors that appear to affect an object or event. When plant growth is observed and considered, the factors of light and water are almost intuitively associated with it. But other factors such as temperature, soil type, and mineral content might also be involved. Isolating these important factors or variables is an essential aspect of interpretation. Discovering that a particular variable is not important represents a significant finding. For example, in working with a pendulum, students will find that the length of the string of the pendulum affects the period, but the weight of the object on the string does not.

The process of interpreting no doubt involves all ten of the rational powers of the mind and is one of the most educationally fruitful aspects of the learning cycle, but it is also one of the most abused parts. Teachers are usually willing to allow children to observe and measure, but when it comes to interpretation, teachers have a tendency to take over. Interpreting data is fun and exciting, but unless the students do it, the educational value is lost. Teachers can act as guides, but they should let students interpret.

There are two procedures for teaching data interpretation—the *individual method* and the *group method*. In the individual method, each student keeps personal records and, with the assistance of the teacher or classmates, interprets, generalizes, and concludes from them. What the students receive from this procedure is directly dependent on them. In the group method, each student contributes personal findings to the entire group, and what the student gains is as much dependent on the data of others as it is on personal data. This latter method, in addition to being useful in demonstrating the value of working together, leads students to find the scientifically sound concept of the value of more than one viewpoint or set of measurements. While there are values to be gained from having students interpret their own data, these values, as well as several others, will also be achieved by group interpretation of data.

In the group approach to interpreting data, the teacher uses the chalkboard and serves as the class secretary. Each group of students states what it found, and the teacher records this. At the end of such a session many data are available for inspection and study. The entire class has the same data, and the attention of the entire group (including the teacher) can be focused on any piece of information. The class is then able to decide whether or not any of the variables being considered require the collection of additional information. This type of social interaction is important in learning and intellectual development.

Data interpretation frequently involves data organization. A collection of unorganized descriptions and measurements usually makes no sense. Imagine a sports page randomly covered with bits of information about the scores of baseball games for the entire season. Before you could begin to discern a pattern, you would almost surely have to decide on some organizational scheme and then implement it. You might begin by deciding to count and record the total number of wins and losses for each team. After that, you might decide to serial-order the teams on the basis of most wins to least wins. The data would then begin to yield new information, which could not be obtained directly from the random pieces of data. The new information might be that Team A has won sixteen of its games while Team D has won only six of its games. The information showing the number of wins or losses by a team could be placed in a histogram, as shown in Figure 4–4. The histogram is just one technique to use in interpreting data. Teachers need to use those techniques that fit the kind of data that are to be organized.

The following learning cycle shows how interpretation was used to lead fifth-grade children to develop the extremely important concept of *community*. The

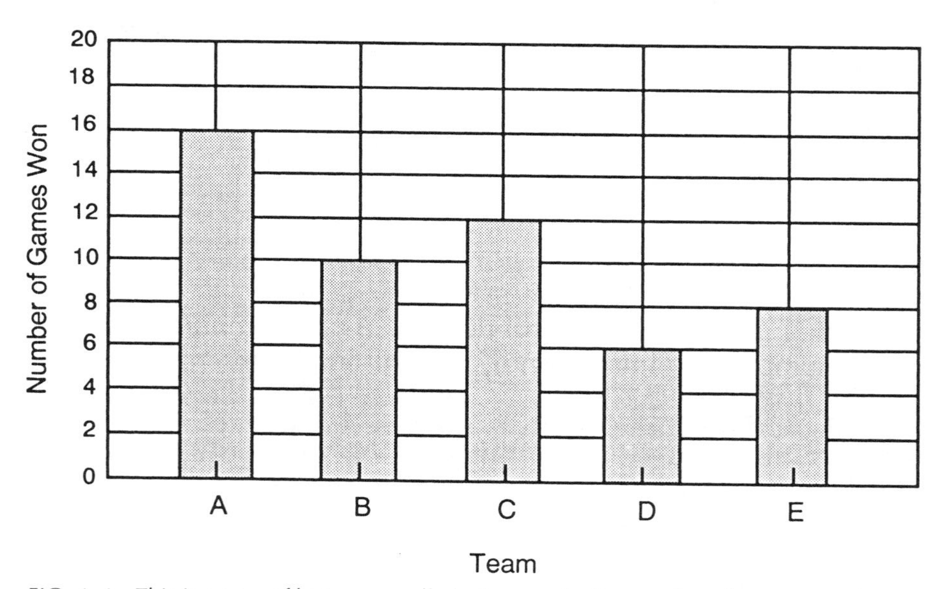

FIG. 4–4 *This is a type of histogram called a bar graph. The number of games won by each team can be readily compared.*

children had collected data for some time about what happens to organisms in small terrariums and aquariums. The information from the exploration phase had been placed on a chart. What follows are instructions for the children to use in interpreting data from the exploration in order to permit them to develop a concept and proceed through the term introduction and concept-application phases. All materials have been written for students.

Exploration

You have been organizing information on a chart in order to represent the relationships among the organisms in a certain area. Now it is possible to look at the chart and to think about each section. You can see how one section is related to other sections.

You should have a section for producers, sections for two kinds of consumers, one section for decomposers, one section for raw materials, and a place to indicate light. Check your work to see if you have included all the sections on the chart. Be sure that you have labeled all the sections.

Choose one section of the chart and imagine that it no longer exists. Imagine that the things represented there have all been destroyed. Write a paragraph about what might happen if that were true.

You have considered what would happen if organisms from an entire section in your chart were gone. Choose just one population. Imagine that all members of that population are gone. How would that affect any other population? Find a population that could be eliminated without an effect on any other population.

Term Introduction

Plants and animals within a certain area are shown on your chart. The chart summarizes the interactions among the living things. The food cycle is an example of interaction among plants and animals.

Organisms in an area interacting in this kind of relationship can be called a

community.

Write the word community at the top of your chart.

The word community is probably familiar to you. You probably live in an area called a community. In some ways, the meaning is exactly the same. The word community refers to the things that are close to you and with which you have some kind of relationship.

The term *biotic community* is applied to organisms that are dependent upon each other for survival. The word *biotic* means "of or having to do with life." A biotic community is a community of living things.

Each part of a biotic community is dependent upon other parts. There are interactions among one living thing and other living things. There are also interactions among living things and the nonliving parts of the environment. One population is dependent upon another population. This dependency holds the community together.

Concept Application

Depending on where you live, you may have several different kinds of communities near you. But, regardless of the kind of community or of the populations that compose it, every community must have the sections that you have represented and labeled on your chart.

Write a paragraph predicting what would happen if you removed "light" from your chart. What would happen to the community? Give data to support what you think.

Think of some other kinds of communities. Choose a kind that interests you and make a community chart that shows different organisms interacting in a food cycle.

Perhaps you will need to investigate organisms and food habits. Be sure that you have all the necessary parts of the community. Draw the arrows that show food transfer. When you are through, trade community charts with a classmate. Check to see that every organism listed in the community has a food source.

When listing organisms on your chart, be sure you list people as part of the community. Discuss how people interact in a community. Compare the effects of people on a community with the effects of other organisms.

Make a list of the ways people affect a community. Indicate whether each way is helpful or harmful. Discuss your list with others. Then change your list if you think you should.

Teaching with the IVP procedure (see Chapter 1) instead of the learning cycle may bring up a problem associated with allowing pupils to interpret data. The pupils might interpret the data incorrectly and arrive at a wrong conclusion. When an incorrect conclusion is reached using the learning cycle, what should you do?

Students must not be allowed to harbor an incorrect concept, but experience and research (Marek and Cavallo, 1995; Marek, Cowan, and Cavallo, 1994) have shown that telling them they are wrong will not allow them to learn the concept correctly. The learners will seem to accept your decisions as an adult authority, but they will probably not begin to disbelieve what their own collected information has indicated and accept what you say. If students are going to learn to classify, compare, evaluate, and use all their rational powers, they must be given experience in classifying, comparing, and evaluating without being made to feel that what they do is not really important because the teacher will decide whether or not their data are correct and interpret those data for them. If applying the rational powers to data produced by an experiment has the adverse effect of leading learners to an erroneous concept or misconception, there is only one way for the students to correct that concept for themselves. They must be provided with the opportunity to apply their rational powers to data from a second experiment that will contradict the first. The students must then decide which evidence is correct, although they have absolutely no basis for making such a decision. The only way they can approach a solution is to repeat both experiments or observations or measurements.

If the data from an experiment lead students to an incorrect concept, then they did something in the experiment improperly. Leading students to see the need to repeat the experiments will also give the class an opportunity to review their procedures. If class-determined procedures are carefully reviewed, the probability of repeating procedural errors (which would again result in learners' arriving at an unacceptable concept) will be reduced. If teachers are to lead learners away from a self-developed concept unacceptable to science, they must make a second experiment available that, although it uses a different route, will provide data to enable the learners to arrive at the acceptable concept.

One of the pervasive concerns of this book is that elementary school science experiences should contribute to the development of the children's rational powers. Data interpretation can make a contribution to that development. Educationally speaking, spending the time needed to make additional observations and measurements and allowing the children, through their own interpretation, to arrive at a correct result is more fruitful than short-circuiting the process by rejecting their interpretation and supplying your own.

Moreover, students can learn about "real" science through their own errors and the need to recheck results. Real science often leads to dead ends and/or includes errors. For these reasons science experiments are repeated and all of the data are carefully checked. Engaging students in such activities places them in the role of scientists—and helps students better understand what science is.

Experimenting

When applied to observation and measurements, data interpretation often leads to other science processes—experimentation, model building, and prediction. The results of using these processes require further interpretation.

When observations and measurements are initially interpreted, one or more possible trends, patterns, or relationships may be noticed or suspected. These possible relationships act as a guide to further explorations and are sometimes called *working hypotheses.* These explorations, which will be designed to establish whether or not a relationship exists, must be carefully planned and controlled. When observations and measurements are made under planned and controlled conditions, an experiment has been done. Since experimenting requires controlled conditions, some data collection and data interpretations should precede it. The data used come from many sources, such as observations, measurements, and the interactions of objects under consideration "to see what happens." Without some data related to the experiment, the investigator would have no idea what to control or what to observe.

Experiments do, of course, provide data, and are an essential part of the *exploration* phase of the learning cycle. Students in a fourth-grade class had some familiarity with plant growth through simple observations. They had decided that certain factors were essential to plant growth and were ready to test them. The following is a series of plant experiments designed for fourth-graders. Notice how certain factors are controlled. Notice also how the directions or guidance given the children decrease as the experiments continue. The directions for students follow:

Exploration

Fill three pots with soil. Plant three bean seeds in each pot. Give each pot the same amount of water.

Set one pot in full light. Turn a box over the second pot. That pot then will be in the dark. Turn a box over the third pot, but make a hole in the box. That pot will then be in a small amount of light.

Each plant will get a different amount of light. Be sure everything else is the same. You are testing only one variable. That variable is the amount of light.

Experiment Repeat the experiment with light. But use grass seeds instead of bean seeds. Put twenty grass seeds in each of the three pots. Keep your records as carefully as you did before. Control all the variables as well as you can. What is the effect of light on grass? What do you conclude?

You have now done controlled experiments with plants and light. Except for light, all the conditions were the same. How does the amount of water a plant receives affect its growth?

Experiment Again, plant beans in three pots. Keep all variables the same except the amount of water. Keep the soil in one pot fairly dry. Keep the soil damp at all times in the second pot. Keep water standing on top of the soil in the third pot.

Observe the growth of the plants. Keep a record of what happens in each pot. Combine your data with the data of others in your class. What do you conclude about water and bean plants?

Experiment Get two small pots the same size. Decide on the kind of plant you would like to grow. You might like to plant sunflower seeds instead of bean seeds. Or you might like to try this experiment with a flower such as the zinnia.

Plant only two seeds in the first pot. Plant a great many seeds in the second pot. Care for the plants in both pots in exactly the same way. Your only variable is the number of seeds you planted in each pot.

Observe the growth of the plants. Keep a record of the growth. What information did you gather from the results of this experiment?

Experiment Next, do an experiment to test the effect of temperature on plants. List the materials you will need. Plan how you will keep track of the results. Write a statement on what you conclude.

At the conclusion of this series of experiments, the idea (concept) *environmental factors* was used to focus on aspects of the environment that affect plant growth. The first two phases of the learning cycle have been completed.

PLAN AT LEAST TWO OR THREE CONCEPT-APPLICATION ACTIVITIES THAT WOULD FOLLOW THE EXPERIMENTS JUST COMPLETED NOW THAT THE EXPLORATION AND TERM INTRODUCTION ARE DONE.

Model Building

Whenever data are interpreted, some type of *explanation* results. Those making the explanation do not know if it describes *exactly* what the system that produced the data is like, but uncertainty is not particularly important at this point in the quest. What is *extremely* important is that the explanation explains the available data. The person producing the explanation has built a mental *model* on the basis of the available data. Observations and possibly measurements will have been made and data—which can come from experimenting—interpreted before model building begins. As soon as an experimenter thinks or says, "those data mean . . .," that student has begun the process of model building.

Quite evidently, preoperational children cannot build models, but early concrete operational children can begin to do so because they are beginning to stabilize the thought reversal process. The following third-grade learning cycle on electricity demonstrates the process of model building. The directions are written for the students.

Exploration

You are going to build an electrical puzzle. You will then exchange your puzzle with someone else. That person will solve your puzzle. You will solve the one you are given. The puzzle will come in a small box. *Do not* open the box.

To make your puzzle, you will need six metal paper fasteners, copper wires, and a small cardboard box. A shoe box will do.

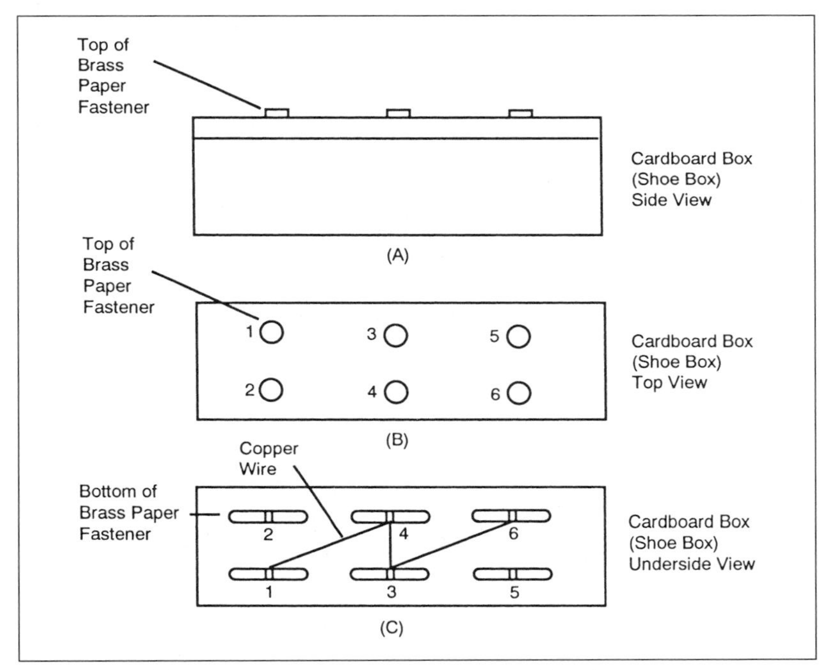

FIG. 4–5 *Different puzzles are constructed by attaching the copper wire to different numbered brass paper fasteners.*

Punch six holes in the box top. Number each hole. Put one paper fastener in each hole. Then connect some of the paper fasteners with wires. Make the connections on the underside of the box top [see Figure 4–5]. You will need a circuit tester [Figure 4–6] to solve the puzzle. Study the directions given [on Figure 4–6]. Those directions will tell you how to make a circuit tester.

Be sure your circuit tester is working well. Touch the ends of the wires to two paper fasteners on your box that are connected. What do you observe? What do the data from your observations tell you?

Now exchange boxes with another group. *Do not* open the box they give you. Touch the tops of different pairs of paper fasteners. Draw a picture of what you found.

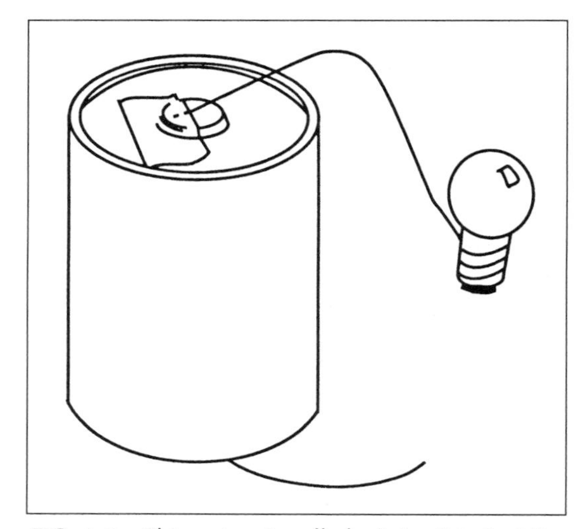

FIG. 4–6 *This system is called a "circuit tester." It is made of a flashlight cell, a bulb, two pieces of wire, and some tape. The bottom of the bulb is placed on one part of an object, while the wire from the bottom of the flashlight cell is touched to another part. If the light goes on, then the object completed the circuit.*

Term Introduction

You found out something from the box you made. You found out something from the box you were given. Write one or two sentences telling what you found.

Concept Application

Use your circuit tester. Go around the room. Touch the ends of the wires to different types of materials. What did you discover?

The children are building a model when they draw the picture of what they believe the inside of the unknown box looks like. Third-grade children *should not* be expected to find all of the connections in the box they are given. To do so they would have to test all possible pairs of the metal paper fasteners two at a time, a problem in combinatorial reasoning that requires formal operational thought. The children will find some connected pairs, and finding all the connected pairs of paper fasteners is not important. What is important is that the students use the data *they* have to draw the picture called for at the end of the exploration phase. When the children who built the box try to correct another group, the teacher should intervene and say something like this: "You know because you built the box. But this group is explaining what their data tell them." All models are based upon the available data, and science contains many such models that have been changed frequently. The solar system and the atom are perhaps the two most well-known models that have undergone frequent changes.

The concept developed here is another model: "The metal paper fasteners must be connected before the bulb lights." The students will probably not say anything about the type of material that must be used to connect the paper fasteners. This information—which is not a model—will emerge from the concept-application phase of the learning cycle. Model building is a very creative process in science. The models of nature created by children or adults enhance understanding or explain some aspect of nature. Furthermore, the model-building process helps students build the quest-for-knowledge model of science and assists them in developing their rational powers *if* the model is based on the available data and *not* on what the answer should be.

RECORD (1) THE AIR TEMPERATURE, (2) CLOUD CONDITION OF THE SKY, (3) THE HUMIDITY, AND THE (4) RATE AND (5) DIRECTION OF THE WIND FOR SEVEN DAYS. ALSO DESCRIBE THE WEATHER EACH DAY. BUILD A MODEL THAT EXPLAINS THE RELATIONSHIPS BETWEEN THE WEATHER AND THE FIVE VARIABLES JUST LISTED.

Predicting

The interpretation of data, as you have seen, can lead to model building. Interpretations and models made from data also have another use. That use is exactly the same as the use that can be made of any past learning in increasing our ability to function more effectively in the future; that is, we use past learning to *predict* what our future behavior should be. The interpretations we make of our past experiences form the basis of our everyday behavior. A common example illustrates this very well. When you are standing at a corner waiting to cross the street and the light regulating the opposing traffic turns red, you feel free (after a visual inspection) to cross. In the past you have had direct experience telling you that moving automobiles stop when confronted by a red light. From that data-interpretation experience, you predict that the moving automobiles you are watching will also stop when the traffic light turns red. You are so convinced your prediction is valid that you are willing to risk your own personal safety by stepping into the street in front of automobiles approaching the red light. This behavior on your part did not come without a great deal of experience, which you accumulated over a long period of time. So it is with predictions that can be made from data derived from experiments. If children are going to learn to predict future events based on scientifically collected and interpreted data, they must be given the freedom to make predictions.

You have seen in this chapter how observing, measuring, interpreting, experimenting, and model building can each provide the focus for exploration, term introduction, and concept application. But where and how does predicting fit into the learning cycle? Before we attempt to answer this question, let us explore in greater depth just what a prediction is. A *prediction* is an estimate of the events to take place, or the results to be achieved, or both.

You will immediately recognize that the description of a prediction does not differ greatly from that of a hypothesis, and your observation is quite correct. There is, however, a fundamental difference between them. A hypothesis is generally based on very limited experience with a particular problem or situation; sometimes it is based on intuition. In other words, a hypothesis is an investigator's belief about what the answer to a question actually is. Although a great amount of evidence is not needed to support the investigator's personal belief, enough is needed to suggest further investigations and guidance for those investigations. Sometimes the information an experiment delivers in its early stages is not definitive enough to allow only one hypothesis to be stated; rather, the data suggest several hypotheses that can be tested. A hypothesis, then, could be described as a tentative assumption stated to enable the investigator to test its validity. Stating hypotheses that are believed to be false is often useful in an investigation because proof of such falsity narrows the number of possible explanations for a problem.

Predictions, however, do not have the tentative, work-guiding nature of hypotheses; they are not stated for the primary purpose of being tested—hypotheses are. A prediction is made on the basis of ideas that have been tested over and over again. For example, weather predictions such as the ones you have done are based on data from such variables as temperature, humidity, time of year, and wind velocity. The effect on weather of each of these factors has been thoroughly investigated, and while that investigation was progressing, many hypotheses about the effects of these factors were tested. Now, however, meteorologists understand the effect of the various factors upon the weather and need not hypothesize about them further. Rather, the effect of such thoroughly tested factors can now be used to predict the weather just as you did. A hypothesis, then, is an assumption that allows the validity of a generalization or model to be tested; a prediction is the use of tested generalizations or models in order to forecast the future behavior of an individual, the results of an experiment, or the outcome of an event. Obviously, predictions involve the use of thought reversals and, therefore, require concrete operational thought.

What is the value to students of learning how to predict? Why is experience in this area an essential part of their experience in science? Perhaps the most basic reason for including experience in predicting in science education is that prediction is a definite, integral part of the structure of the scientific discipline. In many ways, prediction is at the apex of the scientific processes; all that is done in a scientific investigation leads the experimenter toward the goal of stating the results of a similar situation in the future. Since we feel that the elementary school science curriculum must be recognizable as science by a scientist (the integrity of the discipline must be maintained), prediction must be a part of the curriculum.

What is necessary in order to make a prediction? Data must be gathered, classified, compared, analyzed, and evaluated. These data must then be synthesized into a general statement about the situation under consideration, allowing the investigator to reason from the general to the particular about what will happen in a future situation. This, of course, is deduction. In other words, making a prediction demands the use of the learner's rational powers. The experience of predicting, therefore, assists learners in the development of their rational powers and leads them to construct a more complete picture of the structure of the discipline of science than they would if predictions were not included. Prediction is, in a sense, a practical application of one's understanding of nature.

Now to the question, "Where does predicting fit into the learning cycle?" As you can see, predictions are based on ideas that have been formulated primarily through interpretation, experimentation, and model building. The natural place of predicting, therefore, is in concept application. Then why, you might ask, is predicting listed in the "Exploration" triangle (Figure 4–3) as well as in concept application? No idea or concept stands alone. Each idea is linked to others in the structure of science. Even when someone is applying one idea, that person is gathering data with which to construct a related idea or concept. Thus, prediction fits meaningfully into both triangles.

In illustrating the process of prediction in the learning cycle, we will not develop an entire learning cycle. Rather, we include an activity that can be carried out with the concept-application phase of the learning cycle, illustrating model building. This activity is intended to be the last one the student experiences in the learning cycle.

Concept Application

Your teacher will give you a collection of various kinds of material. Write down the name of each type of material. Each material is to be placed between the wires of your circuit tester. *Before you do that,* predict whether or not the bulb will light. Test the material with your circuit tester. Consider only the materials about which you were wrong. Write down why you thought the bulb would light using the material.

Fitting the Essential Science Experiences into School

Preoperational children can make observations and report them, but teachers must remember that these observations will be reported from a very egocentric frame of reference. These children should not be expected to be very objective. They will oftentimes ignore sound reasons obvious to an adult and use such transductive reasoning as "because I like it" or "don't like it" to group objects or make decisions about which plant is greener. Teachers must not come to believe that such children are slow learners or are doing unsatisfactory work in school. These children are just being preoperational. While they can engage in making relative measurements such as bigger than, smaller than, fatter than, longer than, and so on, they cannot use a ruler meaningfully because they do not have the ability to conserve length. But they can do elementary experiments. A child found an odd-looking object among bean seeds and asked, "Is this a seed?" The teacher asked the child what happened to the other seeds they had planted. The child responded that those seeds had "made plants." The teacher led the child to plant the object and when it did not grow the teacher asked the child what she thought. The little girl said, "I guess it wasn't a seed." This is an experiment that late preoperational children can do.

Doing experiments that involve *any* form of record keeping, interpreting data, building simple models, and making even elementary predictions requires that students be capable of making thought reversals. This ability leads them to begin to conserve and give up transductive reasoning. In other words, for children to profit from essential science experiences other than observing, *very simple* experimenting, and interpreting—as we discussed earlier—they must have entered the concrete operational stage. Of extreme importance, however, is providing preoperational children with much experience observing and with other experiences from which they can profit. If preoperational children do not have these experiences with materials, their entry into the concrete operational stage will probably be delayed.

The ability to reverse thinking processes, develop conservation reasoning, and give up transductive reasoning does not appear overnight. The degree of difficulty of the essential experiences from which children can profit becomes greater as the children move more deeply into the stage of concrete reasoning. Most third-graders still do not think like sixth-graders.

The learning cycle has a solid theoretical basis that can guide teachers in their responsibilities because it focuses upon the principal goal of education: learning. The learning cycle was derived from a model of how learning takes place, one that we believe every educator should understand. This understanding—how knowledge is constructed—could and should form the theory base for all education. Educational theory is frequently ridiculed as impractical; however, you have just experienced how practical it is. The theory introduced here can be used to construct curricula to teach students; we contend that this is educationally very practical.

THIS CHAPTER HAS DESCRIBED THE RELATIONSHIPS AMONG THE ELEMENTS OF THE THEORY BASE FOR ELEMENTARY SCHOOL SCIENCE. CONSTRUCT A MODEL OR GRAPHIC THAT REPRESENTS THOSE RELATIONSHIPS. DEVELOP A PAPER-AND-PENCIL MODEL OR CONSTRUCT ONE FROM COMMON MATERIAL OBJECTS.

Other Constructivist Models and the Learning Cycle

The intelligence model of Piaget is the learning and development theory that is the foundation of the learning cycle. Learning and development are very complex and multifaceted processes. Teachers need to be equipped with sound understandings of these processes in order to best promote learning and development among students. We have already discussed how children's learning and development—as defined by the Piagetian model—are promoted and enhanced by using the learning cycle teaching procedure. Other models of learning and development exist in the literature that have application to classroom teaching and learning. These other models of learning and development are also well served through the learning cycle.

With its theoretical basis in Piagetian theory, the learning cycle can be viewed as analogous to the frame of a house. The frame of the house provides the structure upon which the house develops, and the learning cycle provides the means through which the child's intellect may develop. Note that the frame of the house has other features, such as triangular bracing, that provide further stability and improve the development of the house. The carpenter's understanding and use of these features will produce a more sound and refined structure. Likewise, the teacher's understanding and application of other learning and development theories will promote more sound understandings and enhance the intellectual abilities of the child.

Brief summaries of some central tenets of two theories of learning and/or development—the social constructivist theory and meaningful learning theory—are presented below. As you read these sections, think of how these theories are associated with the learning cycle. We encourage you to seek additional knowledge of the philosophies and ideas presented here, as well as those of other constructivist theorists. Such knowledge will provide you with various perspectives and aspects of children's learning and development and thus prepare you for the most effective teaching. Additional literature sources are listed at the end of this chapter.

Social Constructivist Theory

A theory of learning known as *social constructivism* has recently gained great attention for its application to classroom teaching. One reason social constructivism has only recently come to the forefront of education is because this theory originated from the work of a Russian psychologist, Lev Vygotsky (1896–1934). Due to communism in the Soviet Union in the early part of this century, most of Vygotsky's work was banned and thus unknown to psychologists and educators in the western world. Vygotsky's work was banned in the communist Soviet Union primarily because it promoted creative thinking and the development of culturally relevant thought, meanings, and language, which was considered contrary to communist principles. With the fall of communism in the former Soviet Union, Vygotsky's work has become much more widely known to educators around the world. The work of Vygotsky and others who later extended his ideas (e.g., Rogoff, 1990) have many important applications to learning and teaching in schools.

In essence, the social constructivist theory explains that individuals are members of certain cultures that cannot be separated from their learning experiences. Certain skills, ideas, actions, and ways of thinking are valued and promoted by certain cultures—some of which may be common, but some of which are not common among different cultures. The word *culture* may be defined as the ethnic group an individual belongs to, such as Italian-American, or as "subcultures," such as female, Roman Catholic, and science educator. Each of the cultures or subcultures described above implies certain ways of thinking and acting that are familiar and known to others *who are also members* of that culture. There are many cultures and subcultures within society, particularly American society, and we as educators must recognize culture as a significant part of children and their learning and development process. This view of children is precisely the perspective that Vygotsky and others have held in their studies of learning and development.

Within the sociocultural context, Vygotsky described important interactions between learning and development among children. According to Vygotsky (1978):

> It goes without saying that learning as it occurs in the preschool years differs markedly from school learning, which is concerned with the assimilation of the fundamentals of scientific knowledge. But even when, in the period of her first questions, a child assimilates the names of objects in her environment, she is learning. Indeed, can it be doubted that children learn speech from adults; or that, through asking questions and giving answers, children acquire a variety of information; or that, through imitating adults and through being instructed about how to act, children develop an entire repository of skills? Learning and development are interrelated from the child's very first day of life.

Similar to Piaget, Vygotsky believed that learning should be matched in some way to the child's level of development. But Vygotsky did not define specific abilities of learners (e.g., conservation of volume) within four developmental levels as did Piaget. He believed that "we cannot limit ourselves merely to determine developmental levels if we wish to discover the actual relations of the developmental process to learning capabilities" (Vygotsky, 1978).

Instead, Vygotsky believed that educators need to know two levels of development of the child. The first level is the *actual developmental level* of the child. The actual level is described as the completed mental capabilities and functions of the child. In other words, the actual developmental level constitutes what children can mentally do on their own, without assistance from adults or peers. The second level is the *potential developmental level* of the child. The potential developmental level is described by what the child can mentally do with guidance from adults or more capable peers. Once children are able to carry out mental functions at the potential developmental level *without* the guidance of adults or peers, they have attained a "higher" developmental level. This higher level is now their new actual developmental level. Vygotsky noted that it is important for educators to know each child's actual and potential developmental levels, and importantly, the distance between the two levels. He called the distance between the two levels the "zone of proximal development."

> It is the distance between the actual developmental level as determined by independent problem solving and the level of potential development as determined through problem solving under adult guidance or in collaboration with more capable peers. . . . The zone of proximal development defines those functions that have not yet matured but are in the process of maturation, functions that will mature tomorrow but are currently in an embryonic state. (Vygotsky, 1978)

Knowing each child's zone of proximal development is powerful for educators because it allows us to use this dynamic, ever-changing state to help elevate students, mentally, to new levels and capabilities.

It is the educators' responsibility to challenge and help elevate children to new levels of development; this is called *scaffolding*. There are many tools educators can use to help scaffold students, or help them achieve new levels of development. Modeling, questioning, and providing clues (but not answers) are examples of scaffolding tools. However, these tools are only effective if used within students' level of development. For example, if students are developmentally "ready," they can learn to make and interpret line graphs—but not if they have never seen or used one. The students would need to be able to take measurements of two variables (actual developmental level). Given this ability, the teacher may ask students to take measurements of common items around them (e.g., height of their body and hand size). With the teachers' *modeling* and guidance, the students can make a graph and then plot their values on a graph. For example, the teacher would make a model of the *x* axis and *y* axis of the graph for students to imitate. The teacher's probing *questions* will help students begin interpreting the model they created. ("As height increases, what generally seems happen to hand size? How do you know this happens from the information on the graph?") The teacher may provide *clues* to help guide students toward interpreting line graphs ("It is easier to interpret if you read the *x* axis first, followed by the *y* axis.") In this example, the teacher has helped the students think on a higher level than their actual developmental level. Next, the teacher must challenge students to make and interpret graphs of other objects or phenomena on their own, for example, the number of pea plant flowers and the number of pea pods produced by the plant. When the students can make and interpret line graphs on their own, they have reached a new actual developmental level.

In the previous example, the teacher was working within the students' zone of proximal development. The teacher was using scaffolding by modeling graph making. This was accomplished by challenging students with open-ended, probing questions and by giving them clues to help them better read and interpret graphs.

Note also that graph making has a cultural foundation. It is an activity that is valued primarily by more economically and technologically advanced cultures. Graphing is also more valued in certain subcultures than in others. The discipline of science, for example, values and uses graphing data more than the discipline of art. If students tend to have more artistic inclinations and ways of thinking, they may have more difficulty with graphing data than others who, perhaps, come from families of scientists and who value scientific thinking. Thus it is important for teachers to know the child's zone of proximal development and to understand the culturally significant ways of thinking of the child when planning and delivering curricula.

In addition to modeling, questioning, and clues for thinking, there are other initiatives the teacher can take to help scaffold children and move them to higher developmental levels. These initiatives include allowing students to work in groups, to discuss ideas in groups, to display their ideas to others, and to use the language of the discipline (the subculture).

When students work in collaborative or cooperative groups, they can learn from each other. As Vygotsky explained, more knowledgeable or "capable" peers then serve to scaffold other students and help them reach new developmental levels. Together with the teacher, more capable peers can help enhance the overall learning and development of each individual student in the class. In the graphing example, the teacher would set up groups for future investigations such that students who did not master the ability to make and interpret graphs would work with students who had mastered these skills. The more capable peers would help other students in the group make and interpret graphs—and in doing so, help the teacher move all students toward a new level of development.

When students discuss their ideas with others and display their ideas and data (e.g., writing charts, graphs, and explanations on the chalkboard or poster board), they are becoming more adept at the ways of thinking that are valued by the discipline or subculture of science and of scientists. When students use the thought processes (e.g., rational powers), investigative techniques (e.g., predicting), and language that scientists use, *they become scientists* and members of the science subculture. The students can more readily advance to higher and higher developmental levels if the teacher uses all of these tools which help scaffold and challenge students within their zone of proximal development.

THINK OF SOCIAL CONSTRUCTIVIST THEORY IN TERMS OF THE LEARNING CYCLE. HOW ARE THE VARIOUS ASPECTS OF THE

SOCIAL CONSTRUCTIVIST THEORY AND RELATED IDEAS ABOVE EMBEDDED WITHIN THE LEARNING CYCLE?

During the exploration phase of the learning cycle, teachers ask students to model certain procedures (e.g., the use of the microscope), ask probing questions, and provide clues that help scaffold students toward understanding the concept. Teachers also use questioning during the term introduction phase to help students develop, in their own words, their explanations of the concept. Furthermore, in the learning cycle, students work in groups where more knowledgeable students can scaffold other students in the group. Students consistently use the language of the concept (scientific terminology) and communicate their ideas in a variety of forms (charts, graphs, textual explanations, drawings). Students also use materials and equipment that are used by scientists (e.g., a thermometer). By using and practicing (1) science investigations, (2) scientific ways of thinking, (3) scientific language and ways of communicating, and (4) the materials of science, students learn to think more like scientists and become part of the subculture of scientists. These activities are embedded within the learning cycle and help facilitate students' learning of science concepts and development of higher quality of thought.

Meaningful Learning

The theory of *meaningful learning* was described by psychologist David Ausubel (1963). In meaningful learning, children relate new ideas, concepts, and information to what they already know. While learning new ideas and concepts, children actively make mental links between the new ideas and what is known in order that the ideas and concepts make sense and have relevance to them.

The antithesis of meaningful learning is rote learning, in which students memorize content without forming relationships or making sense of the information. The students memorize facts in isolation to one another and do not link ideas. Ideas learned by rote do not make sense and thus do not become relevant to students.

Concepts and ideas that have been learned meaningfully are retained longer than those learned by rote. Furthermore, in meaningful learning children develop a more sound mental network of related ideas that they can *use,* as opposed to lists of facts that are quickly forgotten.

Meaningful learning requires more than "surface" thinking, which would be used, for example, to memorize a random list of numbers (Cavallo, 1991). Meaningful learning involves mentally attaining an interrelated conceptual framework—more closely aligned with describing the *relationship* of each number with other numbers in the list and relating the numbers as a whole to some concept. For example, the numbers in a list may represent changes in values taken on pulse rate in resting, normal, and physically active conditions. The pulse rate changes can be related to other changes in the body, such as heart rate and breathing rate. Measurements that can be related to some object, event, or phenomenon are *mean-*

ingful to students, and thus concepts associated with these measurements are more readily understood and retained.

Science concepts that we plan to help children learn are often based upon or closely linked to previously learned concepts. The previously learned concepts are called *subsumers* because they provide anchorage for newly learned concepts (Ausubel, 1963). One problem with continued rote learning is that students will not have created these anchorage points for linking newly learned science concepts. These anchors are the basis for understanding. In rote learning, newly learned science concepts are not anchored to previously known concepts and consequently do not make sense to the students. Eventually, students will have no choice but to continue memorizing more and more difficult concepts, and they will likely become frustrated and conceptually "lost," and will stop taking science courses in high school (Novak, 1988).

Ausubel (1963) defined three criteria that are necessary for promoting meaningful learning. First, the learner must have relevant prior knowledge. The criterion of "relevant prior knowledge" means that the learner must have a developed framework of understanding (anchors) upon which to link new ideas, concepts, and information. Second, the learner must manifest the meaningful learning set. To fulfill this criterion the learner must actively attempt to relate aspects of new ideas, concepts, information, or situations. This criterion means that the student must be an active part of the learning and must be thinking of ways that newly learned ideas "fit" or connect with other ideas they know. Third, the learner must be given meaningful learning tasks. This implies that the teacher must choose to teach concepts that *can* be meaningfully learned. The teacher must also present learning experiences for students that help facilitate meaningful learning of concepts and avoid rote learning.

Ausubel (1963) recognized the need for young children to have direct experiences with objects and phenomena to acquire meaningful understanding.

> Learners who have not yet developed beyond the concrete stage of cognitive development are unable meaningfully to incorporate within their cognitive structures a relationship between two or more abstractions unless they have the benefit of current or recently prior concrete-empirical experience.

In other words, elementary school children need to manipulate the actual objects and gain direct experiences in order to learn meaningfully.

For example, learning about clouds by copying down their technical names and definitions will be rote, even if the students see pictures in a book. This activity does not use the students' relevant prior knowledge, and the students will not likely make mental links primarily because the task is not a meaningful one.

An alternative activity that would promote meaningful learning would be for students to go outside and draw the clouds they observe in the sky each day over a period of time. At the same time, the teacher could ask the children to record the temperature, time, and other weather conditions (e.g., rain, haze, humidity) for that day. The teacher could post around the room the cloud drawings labeled with the date and time. The children can create names and descriptions for their clouds according to how the clouds appear. For example, they may say that "fluffy, cottonlike clouds" are usually high in the sky on clear, sunny days with low humidity. The students could name all of their clouds and develop descriptions of each. The students could group their cloud drawings together with other students' drawings according to similar features. The students could make a class chart of their cloud drawings and definitions with the "different kinds of clouds" making the rows and "different weather conditions" making the titles of the columns. The teacher could then lead students toward developing the concept that different kinds of clouds are present in the sky when certain other conditions are present (certain temperatures, times of the day, humidity). Thus the students have made links between the cloud types they personally observed and the associated weather conditions. In the future when the students go outside they will see clouds and better understand why that particular cloud is present at a particular time or on a particular day. After the children have made these relationships and links between clouds and other weather conditions, they are ready to open a textbook with pictures and learn the scientific names for their clouds. The teacher and students can post the scientific names beneath the clouds' made-up names. Now the reading of the textbook will be more meaningful because the students' experiences with observing and studying real clouds serve as anchors for new ideas and information. In the above activity, the teacher made use of students' relevant prior knowledge, such as how to take weather measurements (e.g., temperature). The teacher helped students develop new knowledge through concrete experiences, and this knowledge will serve as the relevant prior knowledge needed for future learning about weather. The teacher helped motivate students to actively make links between ideas by *engaging* them in the learning. The teacher provided tasks (activities) that were meaningful to students and relevant in their everyday lives.

Ausubel (1963), and later Novak and Gowin (1984), cautioned that in "discovery" classrooms students may still memorize facts. Such rote learning can occur in discovery-oriented classrooms when the teacher gives too much direction, "gives away" concepts, and does not let students explain *their own* understanding of meanings related to their discoveries. Rote learning can also occur if the students are not given experiences that help them mentally link the newly learned concepts to previously learned concepts and ideas, and to their life experiences.

THINK OF MEANINGFUL LEARNING IN TERMS OF THE LEARNING CYCLE. HOW ARE THE VARIOUS ASPECTS OF MEANINGFUL LEARNING EMBEDDED WITHIN THE LEARNING CYCLE?

The learning cycle promotes meaningful learning because students must construct, formulate, and explain their ideas from their own experiences. The students are not *given answers*, which tend to close their minds and stop their process of making links and meaning of their experiences. Textbook definitions and readings are used by students only after having direct experience with the phenomena. Thus, students first form a knowledge base of understanding of the concept that was central to their concrete experiences in the exploration. This knowledge base is the relevant prior knowledge upon which to link new ideas they learn in the concept-application phase of the learning cycle. Furthermore, the concept-application phase typically includes many activities that help students link ideas and relate them to their everyday lives. You may recall that Piaget labeled the process of linking ideas within the mental structure as "organization."

The learning cycle also provides meaningful learning tasks for students. Students' observations, data gathering, and measurements are based on real objects and phenomena that they have directly experienced. The teacher must carefully plan the whole year's curricula so that the concepts learned in earlier learning cycles can serve as anchors for linking concepts of later learning cycles. Teachers within the school and district should plan a sequence of learning cycles across the grades in which the basic concepts learned in the earlier grades can serve as anchors for more complex concepts learned in later grades. Development and coordination of schools' science curricula to form an interconnected and integrated framework of topics will help promote meaningful learning among students.

References

AUSUBEL, D. P. 1963. *The Psychology of Meaningful Verbal Learning*. New York: Grune and Stratton.

CAVALLO, A. M. L. 1991. "Relationships Between Students' Meaningful Learning Orientation and Their Mental Models of Meiosis and Genetics." *Dissertation Abstracts International*, 52(08), 2877A. (University Microfilms No. AAC92-04496).

EDUCATIONAL POLICIES COMMISSION. 1961. *The Central Purpose of American Education*. Washington, DC: National Education Association.

MAREK, E. A., AND A. M. L. CAVALLO. 1995. "Passkeys to Learning Science in the Elementary Schools: The Data and Language of Science." *Journal of Elementary Science Education* 7(1): 1–15.

MAREK, E. A., C. C. COWAN, AND A. M. L. CAVALLO. 1994. "Students' Misconceptions About Diffusion: How Can They Be Eliminated?" *The American Biology Teacher* 56(2): 74–77.

NOVAK, J. D. 1988. "Learning Science and the Science of Learning." *Studies in Science Education* 15:77–101.

NOVAK, J. D., AND D. B. GOWIN. 1984. *Learning How to Learn*. Cambridge, MA: Harvard University Press.

PIAGET, J. 1964. "Development and Learning." *Journal of Research in Science Teaching* 2(3): 176–186.

PIAGET, J., AND B. INHELDER. 1969. *The Psychology of the Child.* New York: Basic Books.

RENNER, J. W., AND E. A. MAREK. 1990. "An Educational Theory Base for Science Teaching." *Journal of Research in Science Teaching* 27(3): 241–246.

ROGOFF, B. 1990. *Apprenticeship in Thinking: Cognitive Development in Social Context.* New York: Oxford University Press.

VYGOTSKY, L. S. 1978. *Mind in Society: The Development of Higher Psychological Processes.* Cambridge, MA: Harvard University Press.

Additional Resources

AUSUBEL, D. P., AND F. G. ROBINSON. 1969. *School Learning: An Introduction to Educational Psychology.* New York: Holt, Rinehart and Winston.

AUSUBEL, D. P., J. D. NOVAK, AND H. HANESIAN. 1978. *Educational Psychology: A Cognitive View.* 2d Ed. New York: Holt, Rinehart and Winston.

BRUNER, J. 1984. "Vygotsky's Zone of Proximal Development: The Hidden Agenda." In *Children's Learning in the Zone of Proximal Development*, edited by B. Rogoff and J. V. Wertsch, 77–91. San Francisco: Jossey-Bass.

COLLINS, E., AND J. GREEN. 1992. "Learning in Classroom Settings: Making or Breaking a Culture." In *Redefining Student Learning: Roots of Educational Change*, edited by H. H. Marshall, 59–85. Norwood, NJ: Ablex.

GARCIA, E. E. 1993. "Language, Culture and Education." In *Review of Research in Education*, edited by Darling-Hammond. Washington, DC: American Education Research Association.

LURIA, A. R. 1976. *Cognitive Development: Its Cultural and Social Foundations.* Cambridge, MA: Harvard University Press.

MOLL, L. 1990. *Vygotsky and Education: Instructional Implications and Applications of Sociohistorical Psychology.* New York: Cambridge University Press.

NEWMAN, D., P. GRIFFIN, AND M. COLE. 1989. *The Construction Zone: Working for Cognitive Change in School.* New York: Cambridge Press.

NOVAK, J. D. 1977. *A Theory of Education.* Ithaca, New York: Cornell University Press.

VYGOTSKY, L. S. 1986. *Thought and Language.* Cambridge, MA: MIT Press.

WERTSCH, J. V. 1985. *Vygotsky and the Social Formation of the Mind.* Cambridge, MA: Harvard University Press.

———. 1991. *Voices of the Mind: A Sociocultural Approach to Mediated Action.* Cambridge, MA: Harvard University Press.

PART 3

CONCEPT APPLICATION

5 Developing Learning Cycles

How does a teacher go about preparing a learning cycle? How much information should be included in each one? Can other subjects be taught using the learning cycle procedure? There is no single, simple answer to these questions, and no doubt the answers will depend upon whom you ask. The profession of science education has not determined one procedure for preparing a learning cycle—mainly because there are many ways to approach this endeavor. The procedures depend upon many factors ranging from the age levels of the students to the experience levels and preferences of different teachers. Nonetheless, certain guidelines are common to the preparation of all learning cycles. What follows are descriptions of these guidelines and then a discussion of different ways we prepare learning cycles.

Finding What Is to Be Taught

The first guideline, which must be adhered to, is that students in grades two through six are *concrete operational*. Students in kindergarten and most of first grade are *preoperational,* which means they can make observations and report what they *believe* they saw. This guideline is perhaps the most difficult for parents, teachers, and school administrators to accept, because young concrete operational thinkers have such excellent memories. These learners can repeat what we want them to understand and can frequently fool adults into believing that they do understand. Beyond the first grade, students need to be treated as though they reason with concrete operational logic. This leads to a second guideline: *Only concrete concepts can be taught in an elementary school science program.*

These two guidelines demand an answer to the question, What standards need to be used in selecting concrete concepts? *Concrete operational students learn only through direct experience with the phenomenon to be learned.* If the concept that plants need light to survive—as opposed to grow—is to be learned, concrete operational students must grow plants in light and in darkness and collect data from this concrete experience. These data are then used to develop the concept.

To teach science to concrete operational learners, materials and supplies are needed to provide direct experiences. This need leads to an important question: How are concrete concepts differentiated from formal concepts? If materials can provide *direct* experience with a concept, that concept is concrete. This axiom can also be used to distinguish which parts of formal concepts can be taught to concrete operational students. If any part of a formal concept can be taught using materials, *that part* is concrete. For example, diffusion can be taught concretely by having students put several drops of colored dye in a glass container of clear water, observe the results, and verbalize the phenomenon. This is a concrete concept learned through concrete experience. But the concept—diffusion—is formal if taught as the random movement of molecules from a place of higher concentration to a place of lower concentration. Random movement of individual molecules cannot be observed; it requires formal operational thought to be understood.

A formal concept is not made understandable by teaching any part of it that is concrete. The only part that students will understand is the part they directly experience. Furthermore, if the understanding of a concrete portion of the concept depends on first understanding a formal portion, neither portion will be understood. Concrete operational children need to learn science from concrete operational concepts. Preoperational children need to experience science through observation and then discussion.

Deciding What to Do

The Exploration

Assume that you have identified the concept to be taught and the materials to teach it. Remember that elementary school students must have concrete *experiences* with materials to learn concrete science concepts. What does that mean? The children must do experiments and gather data. But before they can do so, the teacher must decide what activities the children will engage in to find the needed data and what records they should keep. In other words, the teacher must prepare instructions for the children to assist them in collecting their information. These instructions need to direct the students' activities, suggest what records students should keep, and *neither tell nor explain* the concept. The following learning cycle, appropriate for fourth grade, illustrates how to present the concept of the *standard unit of measure*.

> Tape two sheets of notebook paper together end-to-end. Place your arm on the sheet with your fingers straight out. Make one pencil mark on the paper at your elbow. Make a second mark at the tip of your middle finger. Use a meter stick and draw a straight line between the two pencil marks.
>
> The line you just drew is the oldest unit of measurement we know about. The Babylonians and Egyptians used this unit many centuries ago. The length of the line you just drew is a *cubit*.

Do not be misled into believing that the introduction of the word "cubit" constitutes a term introduction; it does not. "Cubit" is not a concept. Rather, the word is the name of an object, which is the line just drawn. The exploration phase of the learning cycle continues.

> Measure the length of the longest chalkboard in your classroom. Record that measure. How many cubits long did your classmates find the chalkboard to be? Compare your measurements with the measurements others made. How much do the measurements vary? You and your classmates figure out a way to make a record of those variations. What do you think causes the variations? Make a record of the reasons you and your classmates give.

Notice that the directions given are specific; there is no doubt about how to proceed. No hint is given, however, about what the concept being assimilated is. After the exploration experience, some children make comments such as, "We all need to use the same cubit. That will cut down the variation." These children have assimilated the essence of the concept of a standard measure unit. This is an example, not of *free* assimilation but of *guided* assimilation. We believe that this type of assimilation is appropriate to the constraints on the teaching of science in the elementary school.

A deficiency sometimes found in learning cycles is a lack of experience in the exploration phase. Those preparing learning cycles must remember that the assimilation process *requires time*, and that *time also needs to be spent with the materials and activities* that will lead to the concept. A good exploration provides opportunities for students to assimilate the concept from more than one activity. The following illustrates this point for the standard measuring unit.

> The Roman Empire lasted from 753 B.C. to A.D. 476. During that period the Romans used the length measure called the foot. That measure is based upon the length of the human foot.
>
> Get a piece of cardboard a little longer than your foot. Remove one shoe and sock. Place your foot on the cardboard and stand up. Have a friend place pencil marks at the end of your big toe and at your heel. Use a meter stick to draw a straight line between those marks. The distance between those marks is *one foot.*
>
> Measure the length of your classroom with *your* foot ruler. Record the data from that measurement. Compare your measurement with the measurements of five classmates. Why do you think those measurements varied? Make a record of the reasons you and your classmates gave for the variations found.
>
> The inch had two beginnings. At one time the inch equaled the length of the first joint of the thumb. At another time three barley seeds laid end-to-end were called an inch. Use the first joint of your thumb. Measure the length of a book. Record your measurement. Compare your measurement with the measurements of the same book taken by five others. Make a record of why you think you found variation in the six measurements.

While the exploration phase must provide adequate time for assimilation, it must not go on so long that students get weary. No one can judge that except the teacher in the classroom with the children, and that judgment is crucial. The learning cycle should contain an "extra activity" to be used if, in the teacher's judgment, it is needed. Such an activity could include the following:

> Use beans to measure the length of a book. Make a record of the number of beans you need. Record the numbers of beans used by four others in your class. Compare the number of beans you used with the numbers others used. Discuss with your classmates why the numbers varied. Make a record of why you think variations were found. What could you do to make those variations as small as possible? Make a plan and try it.

These directions are representative of the kinds of instructions that need to be provided for children during the exploration phase of the learning cycle. Do the children need to be given these instructions in writing or can they be given orally? Only the teacher can make that judgment. We have had success with both. But written instructions can provide children with practice in reading and following directions, and probably reduce the misunderstandings that can result from oral directions.

Term Introduction

As you are aware, this phase of the learning cycle is not as student centered as is the exploration phase. The principal role of the teacher in the term introduction phase is to lead the students to accommodation—to construct the scientific concept. Students must focus on the primary findings of the exploration and accommodate *to those findings*. During the students' accommodation, the teacher must introduce the scientific terminology the students will use to refer to their conceptual understandings in the future.

While preparing the learning cycle, the teacher can decide to do the second phase orally and/or in writing. Regardless of which method is used, five factors must be included in a term introduction:

1. The findings of the exploration need to be reviewed and summarized.
2. *All* findings used must be the students'.
3. The concept must be stated in the students' own words.
4. The proper terminology of the concept should be introduced.
5. One or more reasons for the importance of the concept need to be given.

Here is an example of a written term introduction that could be used to guide the concept development discussion:

> You measured the length of the classroom chalkboard with *your* cubit. The number of cubits long you found it to be was different from your classmates. You mea-

sured the length of the classroom in feet. You used the length of your foot. You measured the length of a book in thumb joints. You used your thumb. In each case your measurements were different from those of your classmates. Comparing the measurements of the classroom's length and the book was difficult. Those comparisons were difficult because the lengths of arms, feet, and thumbs are different.

All objects you measure with need to be the same. All cubits need to be the same. The length of the foot needs to be the same. So does the length of an inch. When all units are the same people can talk to each other about them. When someone says, "A foot long," we know what is meant.

Making all cubits or inches or feet the same is making a standard unit.
A group of standard units is a measuring system.

There is nothing magic about a measuring system. For the system to be helpful, everyone must know about it and use it.

Even if the teacher decides to use an oral technique for the term introduction, writing out something like this the first time really helps. It forces the teacher to state what he really believes the students should accommodate to, and helps crystallize the teacher's thinking. If a written term introduction is used, the reading experience—probably aloud—profits the children. A discussion following the term introduction, regardless of the form it takes, is essential.

Concept Application

The purpose of this phase of the learning cycle is to provide the students with the opportunity to *organize* the concept they have just learned with other ideas that relate to it. For preoperational and concrete operational children, such organization requires that they have direct experiences with the concept and the ideas related to it. Of special significance is the principle that *the scientific terminology of the concept must be used during the concept-application phase.* Such use requires the students to continue to accommodate to both the concept *and* the terminology because the terminology leads them to recall the experiences they have had in establishing the concept. The same general guidelines about directions for the students to follow—written or oral—developed for the exploration phase apply equally well to the concept-application phase. If you are unsure about the kinds of directions students need to conduct an investigation, please reread the discussion on preparation of the exploration phase earlier in this chapter. What follows are several activities suggested for use in the concept-application phase of the learning cycle on the standard unit of measure. Notice that the scientific terminology used in these concept-application activities differs from that used in the exploration activities.

Make a standard cubit for your classroom. Each student should now measure the chalkboard's length with the standard cubit. Make a record of your measurement results. Compare your results with those of five classmates. The six of you should write a description of how your results compare.

In the first part of this investigation you measured the chalkboard with your cubit. Compare those results to the results you got when you used the standard cubit. How did the variations in the two sets of measurements change?

Early measuring systems came from nonstandard objects. Grain kernels and body parts were often used. A group of standard units came from those early attempts to measure. Those standard units are the English measuring system. Standard units in the English system are

12 inches = 1 foot
3 feet = 1 yard
5,280 feet = 1 mile
2 pints = 1 quart
16 ounces = 1 pound

Tape pieces of notebook paper together. Use a foot ruler. On the paper, make a square one foot long on each side. You just drew a square foot. Make a square that is one yard on each side. What would you call that? How many square feet are in it? Design a way to find out. How many square feet are in your classroom? Using the standard unit square foot, how could you find out? What other way could you find out?

Get six squares of cardboard. Make each square one standard foot on each side. Make a box by taping the squares together. Each edge of the box should be one standard foot long. You just made a cube. The space inside the box is its *volume*. Your cube has a volume of one standard cubic foot. Explain why that is true. Make a record of your explanation.

The foregoing represents a perfectly adequate application of the concept, and most learning cycles could end here. All disciplines contain content that can be called *culturally imperative*—that is, the content must be taught, not necessarily for the experience provided, but because the culture demands it. Special learning cycles can often be prepared to teach such content, but frequently the concept-application phase of a learning cycle can be used.

The topic of measurement contains content representing a cultural imperative for United States students—the metric system. Students need to become literate in using the standard units in the metric system. Having experienced the learning cycle through the English system, students are now ready to become functional with the content of the metric system. Only a few of the possible activities will be included here to demonstrate how the concept-application phase of the learning cycle can be used to teach the metric system.

Suppose you want to change pounds to ounces. You must multiply the number of pounds by 16. Changing yards to feet means multiplying by 3. Changing feet to inches tells you to multiply by 12. All of those numbers (16, 3, and 12) are different. Then the hard one is changing miles to feet. You have to use 5,280.

Now think about the number 10. If you multiply 10 x 1, what do you get? Next, multiply 10 x 10. What's the answer? Multiplying by 10 is easier than multiplying by 16, 3, 12, or 5,280!

Another measuring system uses the number 10. That system is called the *metric system.* That system has been adopted in Canada, England, and Australia. Only the United States and a few small countries still use the English system.

Get a meter stick. The meter is the standard unit of length in the metric system. Notice that the meter stick has 100 numbers on it. Place one thumb on the number 50. Place your other thumb on the number 51. Notice the distance between your thumbs. That distance is *one* centimeter. There are 100 centimeters in one meter. Study your meter stick. Make a record of why you believe there are 100 centimeters in one meter.

Make a cardboard cube. Make your cube one centimeter wide on each side. The volume of your cube is one *cubic centimeter.* The standard unit of volume in the metric system is one cubic centimeter.

Now imagine that your cubic centimeter is full of water. That much water weighs one *gram.* The gram is the standard unit of weight in the metric system. Food is often sold by the *kilogram.* One kilogram contains 1,000 grams.

Concept-application activities emphasizing the metric system of measurement could continue. We have done very little with weighing in metric units and have not mentioned the liter as a measure of volume in the activities included here. As long as students' interest remains genuine, the activities should continue. If their interest declines, further learning in the metric system will depend on students' using *only* the metric system every time they measure length, weight, or volume. The only way children will develop metric understanding and eventually "think metric" is through constant use of the metric system.

Preparing Learning Cycles

The general guidelines delineated above provide essential knowledge and tools needed to prepare learning cycles. These guidelines should be implemented with the following recommendations in mind:

1. Select the concept students are to learn and write out a concise statement of it.
2. Select the activities students will use to collect data and to guide them toward forming the concept in the term introduction.
3. Prepare instructions to use as an outline and/or to give to students for collecting the data.
4. Be certain that the instructions direct students *only* in the collection of data, and *do not* provide information that allows them to ascertain the concept from the instructions alone.
5. Prepare teacher guidelines for the term introduction phase. The guidelines will consist of carefully structured questions to lead the students *through* the interpretation of their data and *to* the concept.

6. Select the activities to use during the concept-application phase. Be sure these activities freely use the concept *and its* terminology.

7. Prepare evaluation materials that are to be used.

Using these recommendations, learning cycles may be prepared in a variety of written formats that are practical for classroom teaching. The learning cycles we prepare and find useful in teaching are written in the formats of "Teachers' Guides" and, when appropriate, "Students' Guides."

Suggestions for Writing Teachers' Guides

The design and substance of teachers' guides for your learning cycles depend on a variety of factors, such as the experience of individual teachers. In other words, extensive directions may not be useful for experienced teachers. However, when you first begin preparing learning cycles, we urge you to outline them in writing. We find that the longer we teach using the learning cycle, the briefer our outlines become. Principally, we note the concept and the activities in the exploration and concept application. We also prepare a list of questions that will guide students toward formulating the concept.

For teachers—especially those new to the profession—who would find more direction useful in preparing learning cycle guides, a check sheet is presented in Figure 5–1. The check sheet summarizes the guidelines and recommendations for developing learning cycles discussed above.

Suggestions for Preparing Students' Guides

A major decision the teacher must make is whether student instructions will be written or oral. You have no doubt concluded from what we have said here that we prefer providing written directions to the students for all phases of the learning cycle. In our judgment, providing such written directions reduces children's confusion about what they are to do and makes them more independent of the teacher in collecting their data.

For students in the early grades (pre-K to third grade), it may not be appropriate to prepare detailed students' guides. But students of all ages should display their data and present explanations in some form. Students may make drawings or models, or discuss their data orally with the teacher, group members, and the class. In addition, older students (fourth to eighth grade) should learn the important skills of recording data and writing interpretations in an organized and informative manner. Recording and interpreting data in written form helps students clarify their own thoughts and understandings of their experiences. Students' written responses also provide teachers with information that can be used for evaluation.

The format of the students' guide may vary according to the age, skill level, and prior knowledge of the students. A typical, *structured* format is a students' guide that is *identical* to the teachers' guide with the following exceptions: (1) the concept is not written within the guide; (2) the responses to questions, tables, charts,

A Check Sheet for Preparing Learning Cycle Teachers' Guides

Title

☐ Prepare a title for the learning cycle that does not reveal the concept.

Introduction (Optional)

☐ If appropriate, develop an introduction that raises interest about the forthcoming learning cycle. Do not reveal the concept, but *do* use concepts and scientific terminology of *previous* learning cycles.

Exploration

☐ Select one or more experiments that lead students to one concept.

☐ Provide a list of materials, procedures, and safety precautions for conducting each experiment of the exploration.

☐ Prepare drawings of complex arrangements of equipment.

☐ Describe any technical skills needed to perform the procedures (e.g., focusing a microscope).

☐ Prepare graphs, charts, tables, and questions (that will be used in the students' guide) and display expected data and responses.

Term Introduction

☐ Develop questions that incite higher-level thinking and do not produce only yes/no responses.

☐ List questions in a sequence that helps "scaffold" students toward formulating an explanation and identification of the concept.

☐ Compose anticipated responses to all questions.

☐ In a final question, ask students to state the concept.

☐ Write the concept statement followed by a list of relevant scientific terms and phrases.

Concept Application

☐ Describe the nature of the selected concept application technique(s), for example, additional laboratory activities, demonstrations, readings, and computer technologies.

☐ Supply information needed to conduct the application(s), such as materials, procedures, safety precautions, questions, and reading materials.

☐ Observe guidelines for the exploration above when additional experiments are planned.

Evaluation

☐ Prepare evaluation instruments to measure students' conceptual understandings, process skill development, and other learning outcomes.

☐ Provide anticipated student explanations, performances, responses to questions, solutions to problems, and include criteria for measuring students' accomplishments.

FIG. 5–1

and graphs are not included; (3) the teachers' special instructions are omitted; and (4) the section titles (exploration, term introduction, and concept application) may be changed to terms that students can better understand (e.g., gathering data, getting the idea, and applying the idea). This format is considered structured because the students follow an established sequence of procedures and questions; they also write data and interpretations within the guide.

Another format for the students' guide is an *open* format. In an open format, procedures are given orally and/or written on a handout, chalkboard, or transparencies for overhead projection. The students record data, observations, interpretations, and results in a journal or notebook. The information recorded in the journal may vary for each individual, or the teacher may provide general headings under which students report certain data or respond to questions (e.g., data collected, observations, interpretations of data, the idea, discussion, applications).

Developing Learning Cycles from Non–Learning Cycle Materials

Although many teachers use learning cycles in published curricula as discussed in Chapter 1 (e.g., FOSS, SCIIS), it is also common for teachers to develop their own learning cycles by adapting non–learning cycle activities found in textbooks, activity guides, journals, and other sources. By doing so, teachers can tailor science instruction to their students' needs and interests, and use topics that are relevant to particular classroom settings.

Some commercially produced, non–learning cycle activities are good—but inconsistent with the learning cycle theory base. Such activities, therefore, simply require some editing by a creative and knowledgeable teacher. The following italicized statements identify some common deficiencies found in non–learning cycle activities. These statements (deficiencies) are followed by suggestions for revising the activities to be compatible with the learning cycle teaching procedure and its inherent theory base.

The activity begins with an explanation of the concept (verification activity). Move introductory paragraphs to appear *after* the concept has been discovered by students. The introductory paragraphs can be used to initiate discussion in the concept-application phase. Statements that provide students with too much information can be converted to questions. The teacher can also change a verification activity to inquiry by posing a question to students and allowing them to design experiments to find answers to the question (e.g., "Do plants need light to grow? How can we find the answer to this question?"). Refer to Chapter 1 to review IVP activities.

The activity is teacher-centered or a demonstration. If a teacher-centered activity or a demonstration is safe, the *students* can conduct it as an exploration or concept-application activity.

The activity or demonstration is a discrepant event that gives students the impression that science is "magic" and cannot be explained. A discrepant event is a laboratory activity or demonstration that puzzles the students, primarily because the phenomenon they observed was unexpected. Select only those discrepant events that directly relate to specific science concepts. These events may serve as motivating introductions to certain learning cycles, but the explanations for the phenomenon should not be discussed until *after* students have discovered the concept. When the students can explain their data from the exploration, the discrepant event can be performed again—but this time students will be able to explain the concept demonstrated by the event. Alternatively, discrepant events may be first introduced in the concept-application phase. Again, students will be able to explain their observations based on the exploration and their understanding of the concept. Students realize that science is (usually) explainable—and intriguing!

The activity seems to be "hands-on," but there is no concept, point, or specific science learning that will take place during the activity. Be cautious of doing activities just for the sake of keeping children busy—do not select activities with no definite concept. If there is a relevant concept embedded within the activity, use only those parts of the activity that apply directly to students' discovery of that concept. Also beware of activities that are merely cut-and-paste or art projects. If homemade apparatus is needed for science, students can make the models at home, after school, or during other times of the school day when class work has been completed.

The activity embodies many concepts and is not focused. Complex activities with many concepts can be separated into several learning cycles, each with *one* important and central concept. Some of the "additional" concepts may actually be the "expanded" concept, and therefore the focus of the concept-application phase.

The printed questions are inappropriate. Inappropriate questions include those that require low-level recall, solicit yes/no responses, provide too much information, or are vague. Such questions can be rewritten to require greater use of students' thinking skills. Refer to Chapter 6 for suggestions about appropriate questioning strategies for learning cycles.

There are few or no application activities that help students organize the concept and relate it to their own life experiences. One way to promote students' meaningful learning of the concepts is to ask students to explain or investigate how the concept is related to their everyday lives. For example, ask students to explain how the concept applies to phenomena they have observed in their bodies, homes, communities, or in nature. Give students a few days to investigate applications of the concept in everyday life, and ask them to prepare and deliver short presentations to the class. Such activities help students to see that science concepts are not just classroom knowledge—the concept applies and relates to phenomena in the world around them.

Learning Cycles Integrated with Other Subjects in the Curriculum

The learning cycle is an excellent vehicle for integrating other subjects in the curriculum with science. This integration may be accomplished in different ways, to different degrees, and for different purposes. Two examples of using the learning cycle as a foundation for integrating subjects in the curriculum are presented in the following two sections.

Integrating a Learning Cycle in Physiology

The ensuing learning cycle is on a topic in physiology and is geared for sixth-grade students. The learning cycle was developed by a veteran teacher, William Simpson, who has taught science for over ten years in both large and small schools.

Physiology: You Are What You Eat?

The exploration first involves students in finding and recording their resting pulse rates at their necks. The students are then given a twelve-ounce sample of caffeinated soft drink. After students drink their soft drink, they again measure their pulse rates. In the term introduction phase, students chart their pre- and post-soft drink pulse rates on the chalkboard. Students will observe that, in most cases, pulse rates increase after consuming the soft drink. The class average of pre- and post-pulse rates will reveal an increase in pulse rate after the soft drink has been consumed. Through questioning, students should be guided toward formulating this idea:

Pulse rate may be changed by things we consume.

In the concept-application phase, the students find the pulse of two earthworms. One earthworm is then placed in cold water for two minutes while the second earthworm is placed in warm water for two minutes. When the earthworms are removed from the water, the students measure each organism's pulse rate. The students will find that the earthworm in cold water has a much slower pulse rate than the earthworm in the warm water. Through teacher questioning, the students expand the original concept to include this idea:

Pulse rate may be changed by the amount of heat in the organism's surroundings.

In another concept-application activity, the students again measure and record their pulse rates from their necks. The instructor then turns off the lights, plays soft music in the classroom, and asks students to lay their heads down on their desks. After five minutes, the students measure and record their pulse rates. The instructor then asks students to exercise by rotating their arms, jogging in place, or dancing for three minutes. After exercising, the students again measure and record their pulse rates. After posting resting, normal activity, and exercising pulse rates on the chalkboard, the students will observe that their pulse rates decrease after resting and increase after exercising. The students expand the concept further, this time to include this idea:

Pulse rates may be changed by an organism's physical activity.

Other concept-application activities may engage students in reading articles that discuss the effects of other substances, such as nicotine and alcohol, on pulse rate. The teacher may also provide students with readings and audiovisual materials on the effects of hypothermia and heat stroke on pulse rate.

This learning cycle clearly focuses on concepts integral to the subject of science. However, students' learning of important skills and ideas in other subjects can also be fostered. For mathematics, the teacher can use the data students gathered on pulse rates to help them learn how to calculate averages. The teacher can also help students use their data to create bar or line graphs on their pulse rates during resting, normal activity, and exercising conditions. The teacher can integrate social studies in this learning cycle by leading a discussion on social issues related to caffeine and nicotine addiction. For language arts, the teacher can ask students to conduct library research to write and present reports on other factors that affect pulse rate, such as stress, drugs, and alcohol. Preparing reports on such issues will help students develop their written and oral communication skills. The reading materials given to students can serve as tools for developing proficiency at reading expository text. The students can study different kinds of music (e.g., classical, jazz, country, rock) and examine how and why each form of music may produce different effects on an organism's pulse rate. For physical education, additional exercise activities can take place outdoors or in the gymnasium, where students can learn to find their "target" or optimum pulse rates during aerobic exercise. Students can determine which forms of exercise best help them attain their target pulse rates and achieve cardiovascular fitness.

Integrating a Learning Cycle in Physical Science

The following learning cycle on a topic in physical science is adapted from a SCIS activity for the third grade called *Clay Boats: Experiments with Sinking, Floating and Simple Volume Relationships* (Alberti, Davitt, Ferguson, and Repass, 1976); an integrated version of this learning cycle is published as a book chapter in *Behind the Methods Class Door: Educating Elementary and Middle School Science Teachers* (Cavallo and Schafer, 1994).

Physical Science: Sink or Float?

During the exploration, students read aloud and/or listen to a story about a girl shipwrecked on an island in the Pacific Ocean. The story is accompanied by a drawing of "Shipwrecked Shirley" on the overhead projector. The story line uses a female as the main character to promote the view that girls can "do" science and to avoid the stereotype of science as a male domain (Kahle, 1987). The following is a story that can be used to begin the exploration.

> Shirley has no food on her island and she is very hungry and lonely. She can see across the water to another island that has many coconut trees. She decides that she must travel to that island and gather coconuts for food, and then find her way home. Shirley has only clay to take her off the island she is on. Help Shirley make something out of clay that will enable her to go to the other island for food, and then find her way home. She must be very careful because the water is full of many dangers—including sharks!

The students are given a clear plastic container filled with water, a golf-ball-size lump of clay, a large marble to represent Shirley, and several small marbles to represent coconuts that she will take with her. All materials distributed among the groups are of equal size and weight. In groups of three to four individuals, students manipulate the clay to make something that will float in the container of water. The word "boat" is purposely not used during the exploration so students do not focus only on making boatlike designs. During the first try, many students' clay floating objects (canoes, inner tubes) may sink. The teacher circulates among the groups, asking probing questions such as, "Why do you think your clay object did not float? How can you change your design to make the clay float?" The purpose of these questions is to help students think about their designs and generate ideas for improving their clay objects. The teacher should allow enough time so that eventually all groups experience success and are able to make clay objects that float. The students then test the weight of "Shirley" by placing the marble in their clay floating objects, as illustrated in Figure 5–2.

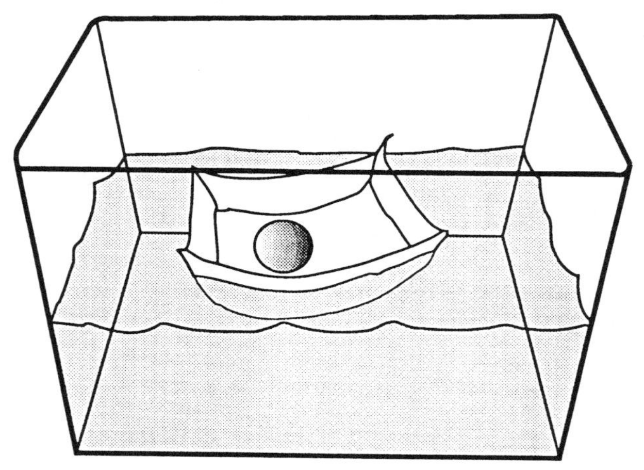

FIG. 5–2

Many clay models may sink with the extra weight. The students then plan and reconfigure their clay to support the weight of the marble. Students draw both sinking and floating designs in their notebooks or on a prepared chart, as shown in Figure 5–3.

After the students' clay structures successfully support the weight of one marble, the teacher informs them that "Now Shirley has arrived at the second island and is gathering coconuts to take with her on her adventure home." Students are given the additional marbles to place in their clay floating objects. The students test how many marbles their floating objects can support. The students compare their clay objects with those of other groups and record the number of marbles supported by the different floating objects. The students may plan and reconfigure their clay floating objects again to support as many marbles as possible.

During the term introduction phase, one member of each group is asked to draw both their floating and their sinking models on the chalkboard. These pic-

Floating Object Drawings	Did this floating object float? Describe.	Did it float with Shirley on board? Describe.	Did it float with Shirley and coconuts? How many coconuts?
Floating Object # 1	No Too thick	No	No
Floating Object # 2	No Too thick Too flat	No	No
Floating Object # 3	Yes Made it thinner and turned up the sides	No The marble made it too heavy on one side	No
Floating Object # 4	Yes Made it thinner all over and turned up the sides even more	Yes	Yes 7

FIG. 5–3

tures and the actual models are observed by all students as the teacher asks questions: "What properties do the sinkers have in common? What properties do the floaters have in common? How are floaters and sinkers different? How do floaters with small bottoms float differently than floaters with large bottoms? When cargo is added to a boat, how does it float differently?" These questions lead the students to an explanation of the concept of this learning cycle, which they discuss within their groups and record in their journals or notebooks. One member writes the group's idea or concept on the chalkboard beneath their drawings. The students write explanations of the concept, such as the following:

> *Clay objects will float on the water if the material is evenly distributed and relatively thin across the surface of the water. If there is excess clay in any one place on the object, the place with more material will float deeper in the water or make the clay object sink.*

This concept is central to the phenomenon called *buoyancy*. Through further comparisons of the different models and the numbers of marbles each of the clay objects was able to hold, the students may also discover the following: (1) Floating objects or boats with larger bottoms float "higher" in the water (i.e., their bottoms are closer to the surface of the water). Boats with smaller bottoms float "lower" in the water and need high sides. (2) The more space a boat takes up (without allowing water to enter), the better it floats and the more cargo (marbles) it can hold before sinking. (3) When weight (more marbles) is added to a boat, the boat sinks deeper in the water.

In concept-application activities, the students relate the concept to other science concepts and to real-life experiences. For example, in discussions and activities students expand the idea of what makes clay float to consider what makes other objects float, including living things. The students may describe how humans float in water: "We can float upright in the water, but our bottoms (feet) are deeper in the water than if we floated in a horizontal position. If we curl up into a ball, we tend to sink." Examples of swimming animals and their characteristics are discussed in class and, if possible, observed in nature.

Another application involves a discussion of what happens to a canoe when it is being loaded for a trip. The students or teacher may make a homemade videotape showing cargo being added to a canoe in water. The more cargo added, the more the canoe sinks down in the water.

In another concept-application activity, students are given a rectangular two-by-four with weights, which can be hung from the middle of the three different surfaces. The students are asked to float the wood in three different orientations: smallest area down, medium area down, and largest area down. The students may first be asked to predict how far down in the water the bottom surface will be when the block is floated in the three different orientations. The same block and weights are used; therefore the weight of the object stays constant for the different orientations. The students will discover that the smaller the bottom surface area, the deeper down in the liquid that area is when the object floats.

For further application, the students are asked to predict the floating behavior of a can of diet soda and a can of regular soda. Their predictions are tested (an aquarium works well) and students observe that the can of diet soda floats, while the can of regular soda sinks. The students are then challenged to explain the observation in terms of weight and size. Size comparisons reveal no difference between the cans. When placed on a scale, the can of diet soda is found to weigh less than the can of regular soda. After reading and discussing the ingredients in the two cans, the students "explain" this difference in weight when they discover that sugar weighs more than an "equivalent sweetness" of nonsugar sweetener.

The above learning cycle offers many opportunities for integration across the curriculum. The integration of other subjects may be accomplished by developing learning cycles that focus on relevant concepts in the different subjects areas, or by developing activities that help students learn and/or practice certain skills. Some subject areas and examples for integration with this learning cycle are listed below.

Mathematics

1. Students can measure and calculate distances between different geographic locations that Shirley visits upon leaving the island in her clay floating object.
2. The teacher can ask students, "How would you measure the space taken up by a clay boat (milliliters of water displaced by sunken boat plus milliliters of water needed to fill up the inside of the boat)? When a boat is placed in water

and floats, the boat takes up some of the space that was taken by the water. In other words, the boat *displaces* some of the water. Figure out a way to determine how much water (in milliliters) is displaced by a boat."

Language Arts

1. Students should be engaged in writing and orally presenting their work throughout this learning cycle. The students write explanations and draw their models in their journals and/or on the chalkboard. The students display their models and present oral explanations to the class and teacher.
2. Students could write a creative story about Shirley's adventures after leaving the island in her clay floating object.
3. Students can write an imaginary documentary describing the marine life Shirley encounters when she leaves the island, the continents and islands she visits, and/or the people and different cultures she experiences (social studies).
4. Students could set up freshwater aquariums and salt water aquariums in the classroom and keep a log or journal of their observations. In addition to technical writing, this activity helps students learn about different water-dwelling plants and animals and their requirements for survival in the two environments.

Geography

1. Students can be given world maps and asked to chart courses for Shirley. In doing so, students become more familiar with the orientation of continents on the globe. The students can also observe locations of certain oceanic and continental geographical structures (ocean trenches, mountain ranges).
2. Pairs of students can be given a plain rubber ball. One student is given a marker and asked to place a black dot any place on the ball. With the partner not looking, the student who made the mark must explain its location on the ball, so the partner can locate the mark. The students will find it cannot be done unless they draw vertical and horizontal lines on the surface of the ball (a grid). This finding leads them to understand the value and necessity of latitude and longitude lines on the globe.

History

1. Students can study the history of boats and boat construction. Students could discover how boats have evolved over time by doing library research, reading, viewing audiovisual materials, and/or visiting a museum.
2. Students could study famous navigators in history and the places and people these individuals encountered.

Reading

1. The students can read trade books or novels about ocean storms, marine life, or famous shipwrecks. One example is the book *Island of the Blue Dolphins* (O'Dell, 1960), which is a story about a girl shipwrecked alone on an island. Other possible books to read as expansions at various grade levels include *The Kettleship Pirates* (Peppe, 1983); *The Island of the Skog* (Kellogg, 1973); *The Voyage of the Dawn Treader* (Lewis, 1952); *Treasure Island* (Stevenson, 1994); *Maps and Globes* (Knowlton, 1985); *Kon-Tiki* (Heyerdahl, 1950); *Moby Dick* (Melville, 1991); *The Swiss Family Robinson* (Wyss, 1992); *Robinson Crusoe* (Defoe, 1992); *The Robinson Crusoe Story* (Green, 1990).

2. The students could read current newspaper or journal articles about ships and submarines, tropical storms, and fish and wildlife preservation campaigns such as "Save the Whales."

Art

1. Students can draw or paint their versions of Shirley traveling in the ocean in her clay floating object.

2. Students could make a life-size papier mâché model of Shirley and her floating object and hang it securely from the ceiling. The classroom becomes the ocean, with the surface of the ocean overhead. The students could then make models of other ocean creatures and treasures and hang or place them at different "depths" within their ocean classroom. While creating these art projects, the students can study characteristics of ocean-dwelling organisms (science) and of other objects found at the bottom of the ocean, such as sunken ships (history).

COMPARE AND CONTRAST THE NATURE OF THE INTEGRATION ACTIVITIES FOR THE LEARNING CYCLES ON PULSE RATE AND BUOYANCY. HOW ARE THESE LEARNING CYCLES AND INTEGRATION PARADIGMS SIMILAR? HOW ARE THEY DIFFERENT?

The learning cycle in physiology centers on a concept that is integral to the discipline of science: factors that affect pulse rate. Note that most integration activities in the other subject areas also center on this same science concept. Thus, the focus of the instruction in the different subjects is to help students gain and expand their understandings of the *science concept*. The physical science learning cycle also centers on a concept central to the discipline of science (buoyancy), but most integration activities center on skills and/or concepts that are integral to the particular subject areas (not necessarily buoyancy). In the physical science learning cycle, the focus of instruction in the different subjects is on skills and concepts that relate in some way to an overarching and common *theme* (e.g., boats or oceans). For additional information on integrating subjects and developing thematic units, we recommend the text *Children Exploring Their World: Theme Teaching in Elementary School* by Sean Walmsley (1994).

These examples illustrate the versatility of the learning cycle for integrating other subjects in the elementary school curriculum. The learning cycle allows students to achieve learning and development outcomes not only in science, as discussed in previous chapters of this textbook, but also in other, diverse subject areas.

SELECT A LEARNING CYCLE FROM CHAPTER 8 OF THIS TEXTBOOK. GENERATE IDEAS FOR INTEGRATING OTHER SUBJECTS IN THE ELEMENTARY SCHOOL CURRICULUM WITH THE SELECTED LEARNING CYCLE. PRESENT YOUR IDEAS FOR THE INTEGRATED LEARNING CYCLE TO PEERS IN THE CLASS. IF POSSIBLE, PRESENT THE INTEGRATED LEARNING CYCLE TO A GROUP OF ELEMENTARY SCHOOL CHILDREN. EVALUATE YOUR EFFORTS.

Developing Learning Cycles in Other Subjects in the Curriculum

The learning cycle teaching procedure is often used as a model for teaching subjects other than science in the elementary school curriculum. The learning outcomes are the same whether the concepts are in math, social studies, art, or any other subject—students develop higher-level thinking skills and sound understandings of the concepts. They are also more motivated to learn. How can the learning cycle serve as the foundation for curricula in other subjects? Following are learning cycles for other subjects in elementary school.

A learning cycle in mathematics must involve students in solving problems using manipulative or concrete objects. A variation of the following learning cycle was developed by Tina McGuffin, a fifth-grade teacher at Bodine Elementary School in Oklahoma City, Oklahoma.

Mathematics: Guessing Games

The class is divided into groups of two or three students, and each group is given one bag of "gummy" candy purchased at a store. The teacher then asks the students to count the candy in their bags. Students count the total number of candies found in their bag and record that number on the chalkboard. After observing the totals, the students notice that there are slight differences in the numbers of candy per bag. The teacher then asks students to use the data on the chalkboard to "guess" how many pieces of candy are in an additional, *unopened* bag of candy. The students use the values to make an "estimate" of the number of candies they think are in the extra bag. Thus, the students construct this concept:

Predicting the numbers of objects based on known numbers of similar objects is making an estimation.

In applying the concept, the teacher asks students how the estimated value in this problem was determined. The students likely selected the number that was "in between" those posted on the board. Through questioning, the teacher can help students discover ways to improve estimations by calculating or using means, medians, and modes.

A second application activity could include counting the different colors of candy in each bag and making estimations of the numbers of each color that will be present in the extra bag of candy. The students can post their values on a chart on the chalkboard with the "colors of candy" making the column headings. Each group records the numbers of each color of candy beneath the appropriate color on the chart, as shown in Figure 5–4.

Candy	Red	Yellow	Green	Orange	Total
Group 1	7	4	10	8	29
Group 2	6	5	9	10	30
Group 3	8	5	10	7	30
Group 4	8	5	11	7	31
Group 5	5	3	12	9	29
Group 6	8	7	9	7	31
Group 7	7	5	10	7	29
Group 8	9	4	9	10	32
Group 9	8	3	10	8	29
Class Sums	66	41	90	73	270
Means	7	5	10	8	30
Medians	8	5	10	8	30
Modes	8	5	10	7	29

FIG. 5–4 *A sample data table for a learning cycle in mathematics. The mean values have been rounded.*

The students can use each group's sums for the different colors of candy to calculate means, medians, and modes for each color. The teacher can ask the students questions about similarities and differences between the values they obtained for the mean, median, and mode for each color of candy, and for the whole bag.

The students can use these calculated values to estimate the amount of each color of candy that will be found within the unopened bag of candy. The students can also make a line and/or bar graph of the class data, with "colors of candy" along the horizontal x axis and "numbers of candy" along the vertical y axis, thus learning how to represent their findings in graph form.

For another application activity, the teacher opens the extra bag and asks students to count the candy to determine how close their estimated value is to the actual value. The students learn to calculate percent error of their estimation by finding the difference between the actual and estimated value, and dividing by one hundred. The students can also count the number of each color of candy in the extra bag and determine the percent error of these estimations. If permitted, the final activity would be to allow students to eat the candy (preferably from an unopened bag), but only after making estimates of the total number and of each color of candy in the bag!

In recent years, learning cycles in mathematics such as the one described above have been more commonly used by teachers. Learning cycles in certain other subject areas are not so commonplace, but are relatively easy to develop and implement. Such learning cycles can be quite ingenious, and create tremendous interest and learning outcomes among children. Two such learning cycles were developed by David Quine, an author and educator. Since 1975, he has served as a classroom teacher, district science director, and school headmaster. "Patterns in Paintings" is a learning cycle for art/music, and "Not All Words Are Alike" is a learning cycle for language arts.

Art and Music: Patterns in Paintings

Divide the class into nine groups. Give each group a set of art prints. Make available three to five paintings by each artist. Prints can be found in libraries, or the complete set is available in *Adventures in Art*, Cornerstone Curriculum Project, 2006 Flat Creek Place, Richardson, Texas (Quine, 1994).

Group 1: Early Christian (first through sixth century)

Group 2: Byzantine and Florentine

Group 3: Early Renaissance: Cimabue and Giotto

Group 4: Renaissance: Leonardo da Vinci

Group 5: Reformation: Rembrandt

Group 6: Romantic: Turner and Cole

Group 7: Impressionist: Monet and Renior

Group 8: Post Impressionist: Cezanne and van Gogh

Group 9: Modern: Picasso and Kandinsky

While it is acceptable to tell the children the names of the painters at this time, do not give the names of the groups (the periods) until the term introduction phase of the learning cycle. Have each group observe and describe each painting in their group. They will note the subject matter and the characteristics.

Encourage them to be looking for similarities and differences in the paintings within their group. Students will write the characteristics on the chalkboard in a timeline sequence from left to right. Attach a single painting, which best characterizes the group, under the corresponding characteristic description on the chalkboard. After observing the descriptions of all nine groups, the students realize that while there are *similarities within* each group, there are major *differences among* groups. The students use these descriptions to classify the paintings into nine groups.

The teacher can now begin the term introduction phase. The students should formulate this concept:

Western civilization, in general, and art in particular have been classified into major periods.

The names of each period are now introduced. These periods are listed above.

In applying the concept, the teacher holds up a new set of art prints one painting at a time. The students describe the work. Then, through comparing and contrasting this new work with the characteristics of each period, the students identify the period into which this new painting is to be placed. A second application activity could include teaching children how to paint according the styles and techniques of each period.

A third application could be correlating music with the art periods. For example, play several musical compositions by J. S. Bach. Ask the children to describe the music as they are listening. Then have them compare and contrast the characteristics of Bach's music with each of the periods of art. Finally, ask them to place Bach into a specific period based upon the characteristics of his works and the characteristics of the period. Learning cycles for music can be found in *Music and Moments with the Masters*, Cornerstone Curriculum Project (Quine, 1995).

Language Arts: Not All Words Are Alike

Give the children a passage from literature to read. Then hand them an overlay sheet. It is to be placed over the passage. Openings have been cut into the overlay so that just the particular parts of speech (for example, all nouns, both common and proper) are now visible. The teacher then asks the students to make a list of these words on three-by-five cards. The students group the cards into similar types of words (for example, persons, places, and things). With the teacher's guidance, students form this concept:

Certain kinds of words name persons, places, or things. These words are called nouns.

In applying the concept, the students are asked to continue listing on three-by-five cards additional nouns from other readings. A second application activity could include classifying the cards into the characteristic of the name of any *one of a class of* persons, places, or things (common nouns) and the names of *particular* persons, places, or things (proper nouns). The expanded concept of common and proper nouns would naturally follow the classification into the two groups. The students should also observe the use of capitalization in conjunction with the use of proper nouns. Students would then be asked to find additional common and proper nouns in additional readings to further apply their understandings of nouns.

Of course, this learning cycle could be done for verbs, prepositions, and other parts of speech. Give it a try! Develop learning cycles for other parts of speech.

DEVELOP A LEARNING CYCLE ON A CONCEPT INTEGRAL TO AN ELEMENTARY SCHOOL SUBJECT OTHER THAN SCIENCE. PRESENT THIS LEARNING CYCLE TO PEERS AND/OR A GROUP OF ELEMENTARY SCHOOL CHILDREN. EVALUATE THE LEARNING OUTCOMES OF THIS LEARNING CYCLE AND DISCUSS YOUR FINDINGS WITH YOUR PEERS IN CLASS.

References

ALBERTI, D., R. J. DAVITT, T. A. FERGUSON, AND S. O. REPASS. 1976. *Clay Boats: Experiments with Sinking, Floating and Simple Volume Relationships*. New York: McGraw-Hill.

Cavallo, A. M. L., and L. E. Schafer. 1994. "Helping Teachers Integrate Science Across the Curriculum Using the Learning Cycle." In *Behind the Methods Class Door: Educating Elementary and Middle School Science Teachers*, edited by L. E. Schafer. Columbus, OH: ERIC Clearinghouse for Science, Mathematics and Environmental Education.

DEFOE, D. 1992. *Robinson Crusoe*. New York: Knopf.

GREEN, M. B. 1990. *The Robinson Crusoe Story*. University Park: Pennsylvania State University.

HEYERDAHL, T. 1950. *Kon-Tiki*. Chicago, IL: Rand McNally.

KAHLE, J. B. 1987. "Images of Science: The Physicist and the Cowboy." In *Gender Issues in Science Education*, edited by B. J. Fraser and G. J. Giddings, 1–11. Australia: Curtin University.

KELLOGG, S. 1973. *The Island of the Skog*. New York: Dial.

KNOWLTON, J. 1985. *Maps and Globes*. New York: Harper and Row.

LEWIS, C. S. 1952. *The Voyage of the Dawn Treader*. New York: Macmillan.

MELVILLE, H. 1991. *Moby Dick*. New York: Random House.

O'DELL, S. 1960. *Island of the Blue Dolphins*. New York: Dell.

PEPPE, R. 1983. *The Kettleship Pirates*. Great Britain: Kestrel.

QUINE, D. 1994. *Adventures in Art*. Richardson, TX: Cornerstone Curriculum Project.

———. 1995. *Music and Moments with the Masters*. Richardson, TX: Cornerstone Curriculum Project.

STEVENSON, R. L. 1994. *Treasure Island*. New York: St. Martin's Press.

WALMSLEY, S. A. 1994. *Children Exploring Their World: Theme Teaching in Elementary School*. Portsmouth, NH: Heinemann.

WYSS, J. D. 1992. *The Swiss Family Robinson*. New York: Bantam.

6 Methods and Technologies Within the Learning Cycle

The learning cycle is not *a method*. It is much greater in scope and philosophy than that. The learning cycle is a teaching procedure which, by its design, allows for many methods of teaching—questioning strategies, demonstrations, and group work, for example. In this chapter we will show how common science teaching methodologies can be used within the phases of the learning cycle. For example, we will discuss questioning strategies in all three phases of the learning cycle. We will also present ways to use demonstrations and technologies within the exploration and concept-application phases of the learning cycle. Finally, we will deal with the issue of safety in the learning cycle classroom, and help teachers learn to preserve the well-being of their students while they engage in interactive learning experiences.

Four exemplary teachers—experienced in teaching with the learning cycle—have prepared the sections of this chapter. Collectively these teachers represent many years of successful teaching, and each is currently teaching science.

Questioning Strategies

(by William T. Fix, physics teacher, Norman High School, Norman, Oklahoma)

Questioning is a vital and frequently used tool in schools, especially in classrooms where teachers employ the learning cycle. Teachers can use questioning for the purposes of guiding discussion and for determining the extent to which students understand the concepts being studied. It is in this context that we will discuss questioning strategies.

Questioning is both a science and an art. The "science" of questioning includes *question types*, the *cognitive level* (or load) of the questions, and the *preplanned placement* of questioning in the lesson. The "art" of questioning relates to the use of questions and questioning strategies to enhance students' understanding—by inviting the students to be actively involved in verbalizing their understandings—and to direct the flow of discussion to achieve the teacher's instructional goals. As with all art forms, skill in questioning comes from repeated use, experimentation with the technical aspects of questioning, and experimentation with timing and placement of questions.

The Science of Questioning

Question Types For the purposes of our discussion we will define five types of questions: *mass*, *controlled*, *solitary*, *voluntary*, and *spontaneous*.

The *mass* question is one, generally requiring a short answer, that all or many of the students will know. Because of the nature of the mass question, its content will be relatively simple. For example, the teacher may say, "I would like all of you to answer together, if you know the answer to the question I am about to ask. How many primary colors are there?" Since there is only one correct answer to this question, all the students who know the answer can be expected to answer in unison.

A *controlled* question could be used to follow up the preceding mass question. It is called a controlled question because the teacher controls which student responds to the question. In a controlled question, the question is asked and the student's name is added to the end of the question, for example: "What are the names of the primary colors, James?"

The *solitary* question is addressed to one student, as is the controlled question, but the difference is that, in the solitary question, the student's name is called at the *beginning* of the question as follows: "Jill, why is green not a primary color?" While the difference may seem slight, the impact of asking one or the other type can be significant. Many people believe that the solitary question should be used sparingly in a class discussion because stating someone's name before asking a question may cause the other students' attention to lapse, since the responder has already been designated.

Voluntary questions are answered by one of several students who volunteer to answer, usually by raising their hands. The question is asked by the teacher, the students raise their hands, and the teacher calls on one student to answer the question. "Who can tell the class what primary colors are mixed to produce the color purple?" Hands are raised and the teacher acknowledges one student as the responder.

The fifth question type discussed here is the *spontaneous* question. A spontaneous question can only be properly used if there is more than one possible answer to the question being asked. If, for example, the teacher asked the question, "What colors of roses have you seen?", there would be many correct color names. The idea behind a spontaneous question is that the teacher asks the question and then allows students to respond at will for an extended period without interruption. During the response time, the teacher would probably be recording the responses of the students.

Cognitive Level The quality of thought that goes into answering a question is what determines the cognitive level of that question. Your study of cognitive development will guide you in selecting the cognitive content to be included in the questions you ask. For example, it would be appropriate to ask a second-grade student to name the colors of a rainbow, but it would not be reasonable to ask that student to explain many details of the process by which the rainbow forms. All of the question types discussed earlier, with the possible exception of the mass question, can be asked at varying cognitive levels appropriate to the students.

Preplanned Placement A vital part of planning a learning cycle is thinking through the lesson and developing a set of questions to guide the students through all three phases of the learning cycle. The key to developing a good set of questions is for you, the teacher, to think through the phases of the learning cycle as though you were actually performing the experiments and learning the concept for the first time. What questions came to your mind when you were performing the experiment? What procedural difficulties did you have with the experiment? Can you recall things about the concept that were confusing to you or difficult for you to understand? How were you able to work through these problems when you were learning the concept? Addressing these questions can help you in developing the questions that you will ask the students during the lesson.

Typically, the questions asked during the exploration phase of the learning cycle will be (1) informal, (2) personally directed to students performing procedures, and (3) related to the students' understanding of the procedures and the data being gathered. If you can observe that some data being gathered are in error, for example, you might ask the student to review the procedure or to tell you how the data were obtained.

The questions asked during the term introduction phase will be ones directed toward helping the entire class understand the concept under study. The teacher will (1) ask students how they would group similar class data together, (2) look for patterns in class data, and (3) offer explanations for data that do not fit the patterns. Following an analysis of the data, students will be asked to formulate a statement that explains the data. This explanation can be done either orally by several students or by having all students write the explanation. These explanations will then be used to develop the concept.

Questions asked during the concept-application phase will be dictated by the nature of the activities used in this phase, whether they are laboratory activities, discussion activities, readings, or other projects. Generally the same guidelines are followed as those used in the exploration phase.

The Art of Questioning

Every answer given to a question provides an opportunity for further interaction between the teacher and the students. The teacher uses questions to engage students in thinking and the teacher's responses to the answers given should be used (1) to keep the students engaged in the thinking process and (2) to give them feedback on the answers they have given. This feedback is referred to as *reinforcement*. By its definition, reinforcement has the function of strengthening the questioning process. Reinforcement takes many forms and includes both verbal responses by the teacher, and nonverbal, "body language" responses.

Verbal reinforcement responses can be either positive or negative. Verbal reinforcement responses can be related to the content of the student's answer or they can be related directly to the student giving the answer (affective reinforcements). During reinforcement the teacher should restate information in the language used by students as much as possible. The following is a sample interaction between a teacher and a group of four students who are in the process of sorting a pile of rocks according to their properties.

Teacher: How did the group decide to sort the rocks, Jennifer?
Jennifer: We sorted the rocks by their colors.
Teacher: You used colors as a property to sort the rocks. What colors of rocks did you find in the pile, Ronny?
Ronny: We found black ones and red ones and white ones and green ones.
Teacher: Jim, can you show me the green ones that Ronny was talking about?
Jim: Well, the ones we called green aren't all green, but they have green streaks in them.
Teacher: So the green ones aren't all green. What other colors are in the green ones?
Gayle: They were actually mostly white, but we wanted to separate them from the other white ones because they were different with the green streaks.
Teacher: That was very good that you all noticed the differences among the white rocks. How else might you sort the rocks, instead of using color?
Jennifer: By size.
Gayle: By roughness or smoothness.
Teacher: All right, Jennifer suggests size, and Gayle suggests roughness or smoothness. Both of those would be good ways to sort the rocks. Can you boys think of another way to sort the rocks?
Jim: Well, some are dull and some are shiny. We could sort them that way.
Teacher: Yes, you could separate the dull ones from the shiny ones. That's a good suggestion.
Ronny: We could hit them with a hammer and sort them by hard or soft.
Teacher: Hard or soft would be a good way to group them, but we might ruin our rocks if we hit them with a hammer. Maybe you can think of a less destructive way of finding out if they are hard or soft rocks. Well, this group has done a good job of thinking about how to sort the rocks!

You can easily imagine the nonverbal reinforcements that might accompany this interaction. For example, the teacher might smile, frown, use hand gestures, or make eye contact when a student is talking.

Questioning strategies and questioning interactions require much planning and effort. The rewards of enhanced learning and personal satisfaction that accompany the effective use of questioning make the planning and efforts worthwhile. What do you think?

Demonstrations

(by John King, science teacher, Highland East Junior High School, Moore, Oklahoma)

Demonstrations can be very powerful instructional tools. Most teachers have demonstrations they use year after year to the amazement and delight of their students. Demonstrations can be used to save time and materials and are often used instead of a full laboratory project when the materials being used are too dangerous for students to handle. There are several purposes for using demonstrations: as a discrepant event to introduce a learning cycle without naming or giving away the concept, as an exploration or a concept application, or as a means of teaching a technical skill.

Discrepant Event

The key to using demonstrations for discrepant events is to avoid giving any background information before you begin. Use this type of demonstration to introduce a unit or to begin a particular learning cycle. For instance, to introduce a learning cycle in middle school about the concept named Bernoulli's principle, suspend an air-filled balloon by a string in front of you. Hold a portable hair dryer above and to the right of the balloon, pointing the dryer toward the balloon at an angle. At this time, ask the class what will happen if you turn on the hair dryer. By doing this, you are exposing your students' preconceived expectations and setting the stage for a teachable moment.

The students will generally say that the balloon will move *away* from the hair dryer. After you receive their responses, turn on the dryer. The faster-moving air from the hair dryer will create a low air pressure area between the balloon and the dryer, causing the balloon to move *toward* the hair dryer. Students are astonished by the unexpected results and their minds are already searching for possible explanations. Be careful not to squelch this curiosity by explaining the process at this point. The teacher should allow students the opportunity to arrive at an explanation for themselves. Now the learning cycle can begin and the students can develop an understanding of how pressure differences operate around them.

When you use a demonstration as a discrepant event, remember the following points: (1) do not give any prior information about the concept you are developing, (2) always ask for the students to predict the results—before you begin the demonstration—to expose their preconceived expectations, and (3) allow the students to develop the concept on their own through the use of an appropriate learning cycle. The discrepant event can be a powerful phenomenon for science teachers to use to introduce a topic.

Exploration or Concept Application

You must first have a clearly defined concept in mind to conduct a successful demonstration during the exploration. For example, using Bernoulli's principle again, the concept that you would want the students to attain would be something like the following: *faster-moving air exerts less pressure.* A learning cycle designed to develop Bernoulli's principle could consist of several demonstrations. The examples given next are neither dangerous nor expensive and the *students* could do the experiments. We are using these simple experiments as *models of how to do other more dangerous and/or expensive experiments as demonstrations.*

For the exploration, place a Ping Pong ball in a funnel. Have a student point the funnel up and blow into the bottom. The ball will not come out of the funnel. In fact, as the student blows harder, the pressure above the ball increases directly with the force of exhalation. Ask the students to develop a hypothesis that would explain the phenomenon.

Place an empty 2-liter bottle on its side with the opening facing the end of a table. Put a small wad of paper in the opening and have a student try to blow the paper into the bottle. The higher pressure in the bottle will push the paper wad toward the student's face. Have your students discuss similarities that exist between the two demonstrations.

Now, perform another demonstration. Before class, drill a quarter-inch hole through a wooden toy block. Cut an index card to the size of the sides with the hole. Push a straight pin through the card and let the card rest on the wooden block with the pin in the hole. Have a student hold the block with the card facing up and try to blow it off through the bottom of the block. The card will be pushed to the surface of the block by the high air pressure above it. Have students look for similarities among all three demonstrations. To arrive at the concept, of course you need to use leading questions to direct your students in their concept development.

After your students have developed the concept during the term introduction phase of the learning cycle, the idea will need to be reinforced, applied, and expanded. Using a demonstration during this phase allows you to check the collective understanding of the class. After developing the idea that faster-moving air creates a low pressure area, lay the balloon that you used earlier on a table and hold the string horizontally. Hold a hair dryer parallel to the plane of the balloon and string and direct the flow of air over the balloon. Have the students predict what will happen. They should deduce that the balloon will rise due to the decrease in air pressure above it. If they are having trouble coming up with this response, remind them of earlier experiments and how the balloon was affected by moving air. You may already realize that these same demonstrations could be used as the exploration of the concept if done *before* the concept-identification phase (term introduction). However, we strongly urge that each student or group of students do the demonstrations, since the necessary materials are readily available and safe.

Teaching a Technical Skill

Possibly the easiest and most common use of the demonstration is for teaching a technical skill. Such demonstrations take the place of exploratory activities with certain kinds of equipment. For instance, a shop teacher wouldn't show a student a band saw and say, "Here it is, figure out how it works." Nor would the science teacher use a set of brand-new microscopes for her sixth-grade class to explore and discover correct operating procedures. One example would be dangerous, the other expensive. In instances such as these, you would use a demonstration to teach the skills of operating complex mechanical equipment.

Of course, these demonstrations do not eliminate student involvement in the learning process. Questioning can be used throughout the demonstration to involve students. Ask them to predict what will happen if you turn a certain knob or flip a certain switch. If they suggest actions that are safe for them and the equipment, allow them to proceed, right or wrong.

Remember, when you are using demonstrations you must invest as much time in planning them as you would in planning other experiments. By using demonstrations you are reducing dangers for the students and costs to your program.

Technology

(*by Joseph A. Green, science teacher, Irving Middle School, Norman, Oklahoma*)

As teachers search for innovative ways to stimulate the interest of their students in science, they must always keep in mind how the innovations fit into the three phases of the learning cycle. For example, the teacher may use computer technology in the exploration phase. Computers and other technology allow the teacher to bring into the classroom things that would otherwise be impossible because of the distance to the source, the cost of the item, or the safety of the children. The teacher, through CDs, laser discs, and on-line services, can have the students explore the vastness of outer space, dive into the deepest oceans, or study dangerous animals. Students can ask experts the "how's" and "why's" of the mysteries of life.

When we use the term *technology*, what do we mean; where does it fit into the learning cycle; and how do we use the various technologies effectively?

VCRs and Video Recorders

The first technology many teachers are exposed to is the VCR and videotapes. In the learning cycle, videos can be used in the concept-application phase to extend understandings to other contexts. The teacher can stop the videotape for discussion of an idea developed in the term introduction phase of the learning cycle.

If your school has a video camera and VCR, you can use these in the exploration phase to gather data. For example, instruct your students to fill balloons with water and predict what will happen to the water inside the balloon when it is burst with a pin. Also have the students predict what will happen to their balloons if they are dropped from various heights. Students then go outside and videotape the "bursting" experiments. Ask the students what they observed and if that matched their predictions. After you have discussed the class observations, you can play the videotape and pause at key points to see if the observations and predictions matched. If your VCR will allow you to advance one frame at a time, the observations can be even better!

Compact Discs and Laser Discs

"How do I use a CD or a laser disc if I have only one computer in my classroom?" This question is asked regularly in many schools across the nation. There are devices that connect a computer to a TV or projecting device—thus involving the entire class in the exploration or concept application—to display the images so that everyone can see. If your school is unable to obtain this equipment, you can use some of the discs that require the students to collect data for the exploration phase of the learning cycle. For example, there are laser discs that allow the students to observe how animals walk. With the laser disc control the teacher can select several different animals and ask the students to make their predictions, then advance the disc one frame at a time and observe how the animals move.

There are also laser discs that demonstrate chemical reactions that the teacher cannot bring into the class but deems important. In this situation the students are able to make predictions and observe the results without being in danger or without having to purchase expensive equipment. Of course such methods should be used sparingly so that the students actually *do* the experiments whenever possible. CDs' and laser discs' titles are quite varied, and it is up to the teacher to see if the activities and information can be incorporated into the exploration or concept-application phases. Of course, be sure to preview CDs and laser discs before using them with your students. Most companies have a thirty-day preview policy.

Computers and Software

The explosion of computer software for the K–12 and especially the K–6 market has been exponential. Although the uses are many and varied, be careful to avoid using this valuable technology only as a tutorial. Computers can often facilitate an essential activity of all learning cycles—gathering data.

Data collection can be one of the most difficult tasks for students of any age, but especially for younger students. For example, students may need to read thermometers or measure length and weight. After students have mastered these skills, the teacher could have the students use special devices connected to computers that allow them to collect various types of data such as temperatures, pH of liquids, pulse rates of students, and many more. Imagine asking your students to predict the pH (acid or base) of various household liquids. Computer technology offers a means of gathering such data, which could otherwise be dangerous. Of course weak acids (vinegar, lemon juice) *could* be used in an experiment, but other stronger acids could be explored through computer technology.

One of the strengths of using groups during learning cycles is the cooperative learning that results. But what do you do if you have a student who has special physical needs? How do you allow for the full participation of all students? There exists special software and hardware for this purpose. For physically impaired students there are voice-activated word processors that allow students to record group answers by simply speaking into a microphone. These students can spell-check, grammar-check, and print the final report without ever touching the keyboard. For visually impaired students there are hardware devices that scan textual material and engage the computer to read back the instructions. The scanning can be done by the teacher before class and activated by the student through the use of a keyboard or voice command. Any student, regardless of physical limitations, can become a full member of a cooperative group while *doing* science.

Today it is quite common to have students for whom English is a second language, or students who do not speak English. How do we involve these students in the cooperative group or in our science activities? Do we translate the text into their language? As strange as it might sound, yes, we do! There are currently on the market scanners with special software that will translate from English to several languages. As the software improves, the translation programs will increase, but at the present time we are limited to a few languages.

While using the learning cycle, teachers often assess students' level of mastery during the phases. Should some students repeat the exploration phase? Do the students *understand* the concept? There is software that allows the teacher to test the students' understandings of the concepts and mastery of processes. This software allows the teacher to have the students who did not properly conduct the exploration phase or who did not understand the concept repeat this phase in a different manner until they have mastered the concept. If a student is ready to move at a more rapid pace, the teacher could have more complex concept applications for this student, thus meeting the needs of each student. Teachers may need to maintain a portfolio of the work of some or all of their students. It is now possible to do this electronically.

On-Line Services

Imagine your students conducting an experiment in your classroom and, at the same time or nearly the same time, sharing their data with students in classrooms around the United States and the world. How exciting it would be for your students to know what an elementary student in Japan or Europe discovered from the same experiment. The diversity of answers could certainly add to their excitement and to the discussions. Your students could use electronic mail to talk to students around the world about their experiments and data interpretations!

Technology is a tool. Nearly every day we have available new technology for the classroom. The main thing we must keep in mind when evaluating the use of the technology is this: Where does the technology fit into the phases of the learning cycle and will it facilitate concept understanding and mastery of scientific processes?

Safety

(*by Sharlene Kleine, science teacher, Irving Middle School, Norman, Oklahoma*)

Teaching with the learning cycle frequently poses special management and safety challenges. The learning cycle involves students in active participation on a daily basis, not just on special "lab days." Therefore, students as well as teachers must always be prepared for anything to happen.

Preparing Your Lessons for Safety

When preparing to teach with the learning cycle it is imperative to anticipate any problems that may occur. As the teacher it is your responsibility to be ready to deal with any problem that may arise. You cannot be overly prepared. Poorly planned activities not only hinder learning but can be dangerous as well. Never have students perform an activity that you have not performed yourself. By completing an activity prior to assigning it, you can anticipate difficulties. List any potential confusing points or dangers and be prepared to deal with them. After students have completed a new activity, list the difficulties that were observed and file them with your lesson plans. When you are ready to teach the same lesson to another group, you will be prepared for those problems.

Some younger students do not have the coordination to handle glassware and/or hazardous materials. Make sure activities are developmentally appropriate for your students. When possible, substitute relatively harmless substances for hazardous ones. For example, if an experiment calls for an acid, use vinegar or lemon juice, or if a base is required, use baking soda. Try to limit student exposure to harmful chemicals.

Preparing Your Classroom for Safety

Make sure that safety equipment, such as a fire extinguisher, fire blanket, eyewash station, and first-aid kit, is present and easily accessible. Examine all safety equipment to ensure it is in working order. First-aid kits should be stocked and portable eyewash stations filled. Check pathways from student areas to safety devices and remove any obstacles. The classroom should have adequate ventilation for any experiments that will emit fumes. A thorough inspection of safety equipment will save valuable time in case of an emergency.

One way to avoid potentially dangerous situations is to limit student movement around the classroom during experiments. By gathering equipment in one area or having it in laboratory tote trays, students will not be wandering around the room in search of equipment. If tote trays are used, number each tray to match table or group numbers. With this system each group uses the same equipment each time, and it is also easier to monitor breakage and loss. If trays are not used, gather only the equipment that students will need that day into one central area. By limiting equipment to only what is needed, you reduce the possibility of inappropriate equipment being used or unauthorized experiments being conducted.

Damaged laboratory equipment is a liability in the classroom. Examine all laboratory equipment. Inspect glassware and dispose of any that is cracked, chipped, or broken. Imperfections in glassware diminish its capacity to withstand heat and may cause it to fail during an experiment. Inspect goggles and repair or dispose of them as necessary. Cracked or loose lenses may allow substances to pass through to the eyes. A head strap that is not secure may allow the goggles to fall off while the student is wearing them.

Inspect all materials to be used in student activities. All potentially hazardous chemicals should be stored in locking cabinets. Students' access to these materials should be prohibited. If a student has access to chemicals and is involved in an accident or act of vandalism, you may be held liable. All flammable materials should be stored in a Flammable Liquid Safety cabinet. These cabinets reduce the possibility of flammable substances being ignited in case of an emergency. Although you may know what chemicals are contained in a cabinet, it is likely that you will not be the only person with access to those chemicals. Many schools store their chemicals and equipment in a central area and all teachers have access to them. Therefore, all chemical containers must be labeled. Chemicals purchased from supply companies come with Materials Safety Data Sheets (MSDS). MSDS contain information on ingredients of the substance, physical and chemical properties, health hazards, reactivity with other substances, and handling and disposal precautions.

Removing any unnecessary items in the classroom reduces the possibility of tripping and falling. Keep the classroom as clean and clutter-free as possible. Items such as electrical and extension cords, excess equipment, scattered books and papers, and excess furniture pose potential problems.

Although you should always discuss classroom rules and safety guidelines with your students, it is also a good idea to have a visual reminder displayed in the room. Post laboratory rules and safety guidelines as well as evacuation plans in a conspicuous area of the classroom. This will help remind students of rules as well as alert visitors to expected behavior.

Preparing Your Students for Safety

Now that you and your classroom are prepared for safety, it is time to prepare students for safety in a learning cycle classroom. For many students laboratory science is a new experience. They must be taught appropriate behavior and proper respect for the laboratory, the equipment, and the people with whom they will be working.

From the first day of school, students need to be aware of the behavioral expectations. Discuss classroom rules with the students and explain why they are important. Students are more likely to follow rules if they know why the rules exist. Some teachers, especially teachers of high grade levels, ask the students to generate the rules of conduct and safety for the classroom. This action allows students to think carefully about safety issues and ways to prevent accidents. Following is a sample set of rules used in a sixth-grade classroom.

1. Use extreme caution every moment during laboratory activities.
2. No horseplay or running in the lab.
3. Report any accidents or injuries to the teacher at once.
4. Do not perform any unauthorized experiments.
5. Safety glasses (goggles) are to be worn when working with chemicals, when working with burners, and when instructed to do so by your teacher.
6. Some chemicals can be dangerous. Never taste, smell, or expose skin/clothing to direct contact with any chemicals unless instructed to do so by your teacher.
7. Leave your area clean for the next class.

Some teachers test students over safety rules. A contract among student, parent, and teacher is also a valuable instrument. A contract not only makes the parent aware of your laboratory classroom, it also reinforces the importance of the rules for the student.

Along with the basic rules for students, there are some basic rules for teachers. Never leave the classroom unsupervised. Accidents can happen in a very short period of time. Do not allow students to perform unauthorized experiments. Even if they are working with relatively harmless substances, it is a dangerous precedent to set—later they may experiment with more dangerous chemicals. Finally, remember, you set the tone for safety in your classroom. Lead by example. If an experiment requires the students to wear goggles, you must also wear them. If students observe you demonstrating safe laboratory practices, they will mimic this behavior.

Since many students may not have had experience doing science experiments, familiarize them with the equipment and how to use it before attempting an actual learning cycle. An excellent way to do this is to conduct a mock experiment using water. Have students fill beakers or other containers with water, measure volume using graduated cylinders, heat water under laboratory conditions, and measure the temperature using a thermometer. This practice gives students the opportunity to become comfortable with equipment and procedures in a relatively harmless environment.

Laboratory work in science classrooms is often done in small groups, whether for cooperative learning reasons or due to limited equipment. Assigning each student specific jobs during experiments helps define a student's role and facilitate the learning process. If, for example, you have a group of four students, you may assign four separate jobs. Make one student responsible for obtaining and returning equipment, while another is responsible for making sure the work area is clean. A third student can make sure directions are read and followed. This student can also be designated as the representative to ask the teacher questions if needed. The fourth student could have the responsibility of recording observations and reporting them to the class. This plan helps ensure that all students participate in the laboratory investigations.

All students are different and there will be some who find maintaining proper laboratory behavior a challenge. While disruptive behavior is distracting in all classes, in a laboratory classroom it can put that student or others at risk. Have an alternative assignment ready for students who cannot behave appropriately in the laboratory. Often, students who have been removed from a laboratory experiment miss the excitement and fun and will learn to control their behavior in the future.

When Problems Occur

Accidents will happen, and on a rare occasion a student (or teacher) will be injured. Most common injuries are minor cuts from broken glassware or burns from improper handling of equipment or materials. Be aware of proper first aid for cuts and burns. Identify any substances the student may have been in contact with and know the appropriate treatment for them. In some cases using the incorrect treatment for a chemical burn can actually make it worse. Treatment procedures for individual chemicals can be found in the MSDS.

For your own protection, it is advisable to prepare a written report of accidents and circumstances surrounding the incident. Sign and date the report, and if possible ask another adult witness and/or the student to do the same. Keep the report in a safe place should a review of the incident become necessary in the future.

Doing science with learning cycles is a challenging and rewarding experience. Maintaining a safe and orderly classroom adds to the enjoyment and enables students and teachers alike to gain more from their learning cycle experiences.

7 *Measuring Students' Progress in a Learning Cycle Program*

Central to our educational system is the evaluation of our students' progress. Students' grades are important, and in the present and evolving educational climate, grades continue to gain greater importance. What types of assessments should be used to measure student achievement? How can students who have experienced science through learning cycles be evaluated? These questions, and many others related to evaluating students, can be answered by maintaining focus on this principle: the assessment techniques must *match* the philosophy, form, and goals of instruction. Assessment techniques used in a learning cycle program, consequently, are based upon the theoretical underpinnings of the learning cycle teaching procedure.

The Learning Cycle and Science Assessment

Throughout this textbook we have delineated the theory base of the learning cycle. We discussed *science processes* that are characteristic of the discipline, such as model building, predicting, experimenting, and inferring. We also maintained that the purpose of schools is to promote students' use of the rational powers in order to help them develop *thinking abilities*. We presented models of development and learning that explain how individuals of all ages form *understandings* of the world. Therefore, in a learning cycle program, we must measure students' progress in developing (1) science process skills, (2) rational thinking abilities, and (3) sound conceptual understandings.

What forms of assessment will furnish information needed to reflect students' accomplishment of these skills and understandings? A common form of assessment used in schools is the conventional test, which is usually characterized by multiple choice and true-or-false questions. Although good conventional tests can be useful, these tests have limited applicability for assessing certain skills and understandings (Shavelson and Baxter, 1992). Furthermore, many conventional evaluation instruments focus on what students *do not* know and *cannot* do. What students know and can do are frequently overlooked or are not measured with conventional assessments. In recent years, the science education community has focused on the development of "alternative assessments" to be used in conjunction with, or to replace, conventional examinations (Jones, 1994; Reichel, 1994). *Alternative assessment* means evaluating students by means other than conventional or standardized tests (Murphy, 1994). Alternative assessments can provide rich information on students' development of science process skills, thinking abilities, and concept understandings.

What are some examples of alternative assessments? How can these alternative assessments be used with the learning cycle? The following sections offer alternative assessments for a learning cycle program.

Alternative Assessments in the Learning Cycle

Alternative assessments measure "authentic" learning because they are carried out *while students are engaged* in learning, as well as at a culminating point of their learning. In the learning cycle, alternative assessments are used in the following ways: (1) each day is viewed as an opportunity to assess students' development of process skills and conceptual understandings, and (2) the teacher uses a myriad of techniques and procedures as indicators of students' achievements.

Science Process Assessments

Science process assessments engage students in using their reasoning and logic to respond to a set of questions. Science process assessments are particularly suitable for the learning cycle because students must perform activities before they will be able to answer the questions. Science process questions should require certain responses, but also evaluate whether students know *how* to find their answers. This latter objective is extremely important in a learning cycle program. The following questions are examples of science process questions that could be used for evaluation after students have completed the learning cycle on *standard units* described in Chapter 5. Some of these questions can be given to students informally during a laboratory activity, or more formally in a laboratory practical examination (to be discussed later). Many of the questions can be given to students as a paper-and-pencil examination.

Science Process Questions

1. Use a balance and weigh some water in grams. Weigh 20, 30, 50, and 100 milliliters. How much does one milliliter of water weigh?

2. Which is the longest distance, three meters or three yards? What is your evidence?

3. Ten members of a fourth-grade class each measured the length of a chalkboard with a ten-centimeter ruler. Ten other members of the same class each

measured the length of the same chalkboard with a meter stick. Which group's data do you think would vary the most? Explain why you think so.

4. How many centimeters are there in a square meter? Be sure to explain how you got your answer.

5. Get a meter stick. How many centimeters are in a meter? How many millimeters are there in a centimeter? How many millimeters are in a meter?

6. You measured a cubit using your arm. The class then agreed upon a standard cubit. Why is the standard cubit probably different from your cubit?

7. Why is the length of three barley seeds *not* a good standard unit for the inch?

8. Why is a standard unit of length important in a measuring system?

These questions differ from conventional test items because students need to explain their responses. Multiple choice questions, for example, do not provide the students an opportunity to explain their responses, so the teacher cannot determine the thinking processes students use to solve the problems. Science process questions, as shown here, ask the students to use higher-level thinking abilities to solve the problems, *and* to explain the reasoning behind their solutions.

Illustrations, Physical Models, and Analogies

Illustrations, physical models, and analogies are forms of assessment that can be used to reveal students' views of how certain processes occur, or how certain scientific ideas may be interrelated. Making drawings, three-dimensional models, and analogies may help students better understand complex science concepts because physical representations are more concrete than verbal representations.

Very young students, who are just learning to write words, can use drawings to represent their understandings. For example, if students are engaged in a learning cycle on animal homes (see Appendix, Part A), the teacher can ask them to draw pictures of places where animals live. The teacher will gain information on students' understandings related to where different animals make their homes by examining the drawings and questioning the children about their illustrations.

Students can also make three-dimensional models to represent their understandings of science concepts. The models can be made with everyday materials that students bring from home, such as modeling clay, buttons, coffee cans, and Popsicle sticks. Making models at home also presents opportunities for parents to be involved in their children's school projects. When completed, the students can explain their models to the class, or present them as science fair projects to a larger audience.

Students' understandings of science concepts and topics can also be assessed through analogies. Students are asked to create an analogy for a scientific concept and to explain how aspects of the analogy correspond to that concept. For example, the students may create an analogy for the human body by describing and/or building a model of a factory, school, or car. The students should explain specifically how the parts of a factory, for example, are comparable to the various structures and organs in the human body. For example, how is the human circulatory system like the conveyer belts of a factory? How is gasoline for a car comparable to food for the human body, and why? The students should also explain how the analogy *differs* from the actual structure to ensure they can distinguish between the model and the actual science phenomenon.

Oral Tests and Interviews

Children of all ages love to explain their work and express ideas orally on a one-to-one basis. An oral test or quiz allows students to express their ideas freely, and in doing so, in-depth understandings and thought processes may be revealed. Furthermore, the teacher may discover misconceptions students may hold. Knowledge of students' misconceptions is valuable data when planning science lessons (Marek and Bryant, 1991; Marek and Methven, 1991).

Teachers often discover that students who have difficulty with written tests do well with oral tests. The students' writing fluency does not pose a potential constraint on their ability to express their knowledge. For example, using the children's illustrations of animal homes discussed earlier, the teacher can gain knowledge of students' conceptual understandings by asking them to explain their drawings. For the learning cycle on buoyancy (presented in Chapter 5), the teacher can gain information on students' reasoning and conceptual understandings by asking them to explain the changes they made to the clay to make their objects float. To reveal students' ability to apply their understandings, the teacher can interject scenarios or problems for them to solve using their drawings or models. For example, the teacher could ask students to explain what would happen if their clay objects were placed in larger or smaller bodies of water. The students could test their models in different amounts of water and discuss their observations with the teacher.

The learning cycle provides many opportunities to administer oral tests or interviews to students. The teacher may designate certain days and times to quiz students orally on process skills, concepts, or laboratory procedures that pertain to specific learning cycles or, more generally, the discipline of science. Another option is to prepare a check sheet consisting of the students' names and one or two questions that are relevant to the learning cycle. Using these questions, the teacher can conduct "on the spot" interviews with students *while* they are performing laboratory investigations.

Mental Models

Mental models are open-ended tests in which students are told to "write everything they know" about a given topic (Cavallo, 1994, 1996; Fleener and Marek, 1992). Students' written explanations provide information on the understandings and misunderstandings they may have about a concept. A mental-model test may be given before a learning cycle to achieve a baseline of information on the students' prior knowledge, and again at the conclusion of the learning cycle to examine how their understandings evolved as a result of their experiences.

In administering mental-model tests, students must be given ample time and plenty of paper. The students must be encouraged to write *everything* they know about the topic, and told that they will not be penalized for "incorrect" information. The purpose of a mental-model test is to allow students to represent, on paper, exactly what they have mentally stored about a topic. Teachers need to use any *misinformation* in students' essays for planning future instruction. The extent of accurate information expressed in the students' essays at the conclusion of the learning cycle may be scored and used toward their grade.

Scoring mental-model essays can be done by reviewing all of the essays and placing responses that are similar to each other into the same groups. The teacher can develop a scoring rubric or hierarchy based on the information represented in the responses of the different groups. In developing the rubric, the teacher can analyze both procedural knowledge (*how* some phenomenon occurs) and declarative or conceptual knowledge (*what* happens during some phenomenon, and why it happens the way it does) (Bryant, 1992). The quantity and quality of procedural and conceptual knowledge students express in their responses can be used to indicate their meaningful understanding of the topic (see Cavallo, 1996). The different groups, containing similar responses, may be rank-ordered and assigned point values ranging from zero—for completely incorrect or irrelevant responses—to a point value representing the most thorough and complete responses (see Price and Hein, 1994).

Using the learning cycle presented in Chapter 5 on a physiology concept, for example, the students could be asked to write everything they know about pulse rate. The teacher would develop a scoring rubric based on students' explanations of how pulse rate is affected by things we consume (procedural knowledge). The teacher can also assess students' understanding of what pulse rate *is*, and why it is affected by factors such as caffeine, nicotine, and exercise (conceptual knowledge).

Creative Writing

Creative writing using science concepts and processes provides a unique mechanism for students to express their understandings. Students who are artistically inclined may have special strength in expressing their understandings in a creative manner. For example, after engaging in a learning cycle in which students discovered that the three states of matter are solid, liquid, and gas (see Chapter 8), they could write creative stories imagining *themselves* to be a solid changing to a liquid, and then to a gas. The children could describe what the changes of state might feel like, and portray events that would cause the changes to occur. The children could also depict events that would cause their gas forms to change back to liquid, and liquid forms to change to solid. The teacher could use the students' stories to evaluate science knowledge and understanding, as well as skill development in language arts. The students could also choose the "best" story and act it out in class. Consider how these activities would help students develop the rational power of imagination!

Laboratory Practicals

Since much science learning in the learning cycle takes place while students conduct laboratory experiments, one valuable technique for measuring specific kinds of accomplishments is the laboratory practical. Laboratory practical examinations usually require students to perform skills or techniques, solve problems, or respond to questions using laboratory equipment and materials. The laboratory practical is usually set up by placing specific materials on different tables around the room, and labeling the tables as stations (e.g., Station 1, Station 2). The students are given a set of questions that correspond to each laboratory station. For example, question 1 can be answered only by manipulating the apparatus at Station 1. The students are assigned to start at different stations, and when the teacher gives the signal, the students begin working on the specified question(s) at their stations. A time limit is usually given for students to perform the task and respond to the question at each station (e.g., five minutes). At the teacher's signal, students move to the next sequential station, and so on, until each student has responded to the questions at all stations.

The following is an example of a laboratory practical for eighth-grade physical science. The questions can be taped to the surface of the table at the station and/or printed on the students' examination paper.

Laboratory Practical Examination

Station 1: (*Materials include a triple beam balance and a rock.*) What is the mass of the object?

Station 2: (*Materials include a graduated cylinder, water, and an irregularly shaped rock.*) Determine the volume of the object.

Station 3: (*Materials include a triple beam balance, a ruler, a graduated cylinder, water, and a rectangular block.*) Calculate the density of the object.

Station 4: (*Materials include a triple beam balance, a ruler, a graduated cylinder, water, and an irregularly shaped rock.*) Calculate the density of the object.

Station 5: (*Materials include equal-sized ice cubes, metal spoon, balance, and two beakers with equal volumes of liquid in beaker A and beaker B. Beaker A contains water and beaker B contains isopropyl alcohol.*) Observe the liquids in beakers A and B. Drop one ice cube in beaker A and one in beaker B. Explain the phenomenon you observe in terms of density. Use calculations of density to support your explanation. Remove and discard ice cubes in the sink when the question has been completed.

Note that the laboratory practical engages students in manipulating materials, solving problems, using rational powers, and explaining concepts. Students often like this test format because they are mobile, and can check their answers by repeating their measurements and calculations.

Concept Maps

Concept maps are a source of information on how students form relationships among science concepts and information. For related information, see Novak and Gowin (1984). Concept mapping promotes meaningful learning because students must link concepts and information as they construct the maps. The maps that students create demonstrate their knowledge of how concepts are interrelated and relevant to each other.

In making concept maps, students are asked to arrange concepts in a hierarchical format, and show how each of the concepts are related through interconnecting action words. In scoring concept maps, teachers may analyze the students' selection of concept words, their hierarchical placement of the concept words, and the phrases used to link concepts on the map. Figure 7–1 shows an example of a concept map on flowers.

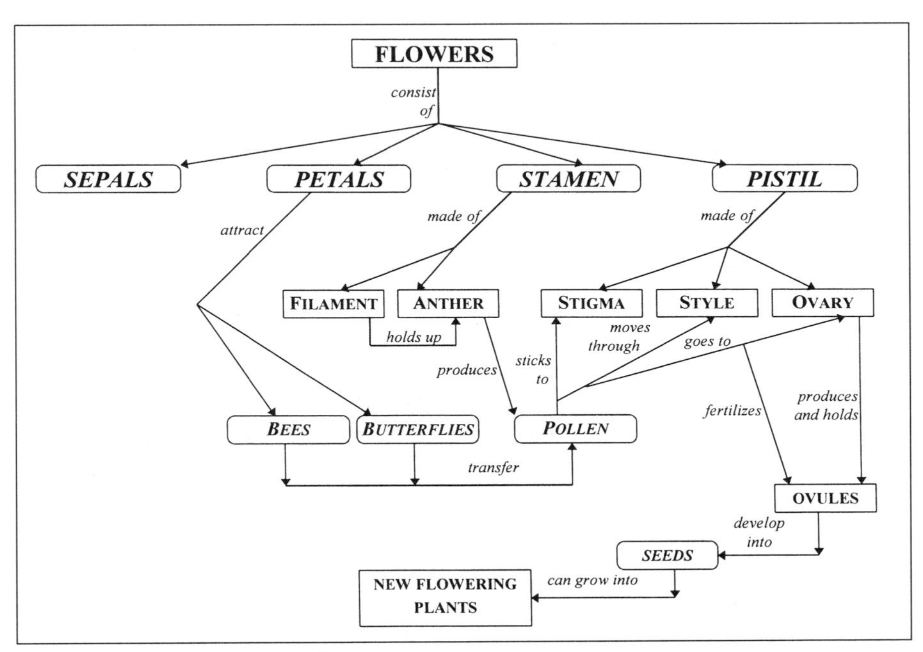

FIG. 7–1 *A concept map on flowers*

During a learning cycle, concept maps can be made by students at the beginning of the exploration to assess prior knowledge and to reveal misconceptions. The concept maps may be restructured or reconstructed by students following the term introduction phase of the learning cycle as they accommodate the concept, and again after concept application while they organize their new understandings. The maps can be made individually, or students may work in groups. Concept maps may be made on large poster paper and displayed around the classroom. The students can explain the organization of their maps to other students and compare and contrast the logic of different organizational structures. Concept maps can be collected and scored as part of a science grade.

Laboratory Journals and Learning Logs

Laboratory journals reveal students' ability to organize and present information about their investigations, such as procedures, data, and conclusions. Journals also show students' logic and reasoning through their drawings, explanations, and resolutions to problems related to the investigations. The laboratory journal is a common record-keeping method that scientists use in research. Thus, the students are practicing techniques scientists use to communicate their explorations and findings, as well as learning some important organizational and technical writing skills.

The journal may be a three-ring notebook that includes students' guides (discussed in Chapter 5), lined writing paper, plain drawing paper, and/or graph paper. Students complete the procedures and questions in the students' guides, prepare explanations, carry out calculations, and construct graphs as directed. In contrast, the laboratory journal may be a simple spiral notebook, in which students develop their own structure for presenting their investigative work. The students may make subheadings for each step of their laboratory investigations (e.g., Exploration, Research Questions, Procedures, Results, Term Introduction, Conclusions, Concept Application) and complete these sections as they progress through a learning cycle. The journal may be checked at the completion of each learning cycle, or it may be turned in for a grade periodically throughout the school year.

Learning logs provide another medium for students to express their knowledge of concepts in a learning cycle program. Students can keep notes in their learning logs every day, or only when engaged in specific experiments or projects. The teacher may provide questions or topics to guide the students, but these topics are usually very broad. For example, the students may be asked to respond to the following questions in their learning logs during a learning cycle:

Learning Log Questions

Before this investigation I thought that. . . .
Today, I learned that. . . .
I will use what I learned today when. . . .

Individual or Group Projects and Presentations

Individual or group projects and presentations are excellent opportunities for students to expand their understandings of science topics. The students can conduct independent experiments, library research, and on-line computer research to learn more about a concept they discovered in a learning cycle. For example, after completing several learning cycles on ecosystems, the students may select a specific ecological problem, such as deforestation, or chose an endangered species in an ecosystem, such as the Florida manatee or Oklahoma lest tern. Working in groups, the students could prepare presentations on their selected topic, highlighting the impact of the problem on ecosystems. The students can communicate their findings in poster exhibitions or oral presentations, similar to the style and format used by scientists at professional conferences. The teacher and students may invite other classes, teachers, and parents to the presentations to showcase the work students have done in science.

To help encourage students' understandings of the topics presented by other groups in class, they may be asked to (1) write questions for the presenters, or (2) respond to open-ended questions the presenters formulate about the topics. The questions and responses can be collected and analyzed for a portion of the students' grades for the projects.

Teacher Observations

The teacher may wish to use informal observations of students as a component of their science grade. In doing so, the teacher may develop an observation check sheet for noting any special improvements or efforts students demonstrate in class. For example, if a student asks a particularly thoughtful question or solves a difficult problem, the teacher would write a note next to that student's name on the check sheet. Periodically during the school year, the teacher may compile information from the check sheets to send notes home to parents, use as report card comments, or to factor into the students' grades.

Portfolios

Portfolios are collections of the students' work over a unit of study, grading period, or the school year (Collins, 1992). A portfolio may consist of a variety of conventional and alternative assessments, as well as nongraded or extra-credit work. In a portfolio, students may place samples of their projects, concept maps, presentation notes, journals, creative stories, and any other relevant work. The portfolio presents the student's best work in science for a given period of time (e.g., the school year). Portfolios can be used to measure a wide range of students' learning and achievements in a learning cycle program. Sample portfolios can be maintained as a chronicle of the different science activities that took place in the classroom over the school year.

Conventional Tests

As mentioned, most conventional tests have limited ability to measure certain skills, thinking abilities, and understandings students have accomplished, especially when compared to alternative assessments. Nonetheless, if properly constructed and clearly focused, these assessments can be useful to assess the rational power of recall. If the teacher wishes to test students on vocabulary words and definitions, for example, short conventional tests would be appropriate. Conventional tests on problem solving must be written to challenge students to think at higher levels of reasoning and allow them to show or explain their work.

Since most standardized tests use conventional formats, it is important that students have some experience responding to questions written in these formats. Experience with conventional tests will prepare students to take standardized tests and other examinations in secondary school and college. Teachers are cautioned, however, to look for or develop test questions that use higher-order rational powers (e.g., comparing, deducing, generalizing) *and not* to overuse conventional test formats to measure students' progress. In other words, teachers *may* use conventional tests in conjunction with a variety of alternative assessments.

Alternative assessments can be implemented in a learning cycle program for students of all grade levels. Children in the early grades can draw pictures of their investigations—such as planting beans, counting the number of beans that were produced from one seed, or mixing certain liquids to discover what colors appear. Students can begin to use "invented spelling" (writing words as they sound to them) as soon as they are able to write, and the teacher can use the opportunity to help them correct their spelling and learn new vocabulary words. In later grades, students can begin to write simple sentences to augment their drawings. The presentations students do can take the form of "show and tell" in the lower grades. Oral tests and interviews are particularly useful with younger students, because they are better able to express their understandings orally than in writing.

Older students need the practice of communicating their ideas and thoughts in a variety of forms. The different formats students use to express their knowledge are useful to the teacher for assessment purposes *and* they are important to the students because they develop skills of representing and communicating their ideas in ways that are useful for everyday life.

SELECT ONE OR MORE OF THE LEARNING CYCLES PRESENTED IN CHAPTER 8 AND APPENDIX A. GENERATE IDEAS FOR ASSESSING STUDENTS' SCIENCE PROCESS SKILLS, RATIONAL THINKING ABILITIES, AND CONCEPTUAL UNDERSTANDINGS AS THEY PROGRESS THROUGH THE SELECTED LEARNING CYCLE(S). PRESENT YOUR IDEAS TO THE CLASS.

References

BRYANT, R. J. 1992. *Students' Declarative and Procedural Knowledge of Heat and Temperature*. Unpublished doctoral dissertation. University of Oklahoma, Norman, OK.

CAVALLO, A. M. L. 1996. "Meaningful Learning, Reasoning Ability, and Students' Understanding and Problem Solving of Topics in Genetics." *Journal of Research in Science Teaching* 33(6): 625–656.

———. 1994. "Do Females Learn Biological Topics by Rote More Than Males?" *American Biology Teacher* 56(6): 348–352.

COLLINS, A. 1992. "Portfolios for Science Education: Issues in Purpose, Structure, and Authenticity." *Science Education* 76(4): 451–463.

FLEENER, J., AND E. A. MAREK. 1992. "Testing in the Learning Cycle." *Science Scope* 15(6): 48–49.

JONES, M. G. 1994. "Assessment Potpourri." *Science and Children* 31(2): 14–17.

MAREK, E. A., AND R. J. BRYANT. 1991. "On Research: Your Teaching Methods May Influence Your Students' Understanding of Common Science Concepts." *Science Scope* 14(4): 44–45, 60.

MAREK, E. A., AND S. B. METHVEN. 1991. "Effects of the Learning Cycle Upon Student and Classroom Teacher Performance." *Journal of Research in Science Teaching* 28(1): 41–53.

MURPHY, N. 1994. "Helping Preservice Teachers Master Authentic Assessment for the Learning Cycle Model." In *Behind the Methods Class Door: Educating Elementary and Middle School Science Teachers*, edited by L. E. Schafer, 13–31. Columbus, OH: ERIC Clearinghouse for Science, Mathematics and Environmental Education.

NOVAK , J. D., AND D. B. GOWIN. 1984. *Learning How to Learn*. New York: Cambridge University Press.

PRICE, S., AND G. E. HEIN. 1994. "Scoring Active Assessments." *Science and Children* 31(2): 26–29.

REICHEL, A. G. 1994. "Performance Assessment: Five Practical Approaches." *Science and Children* 31(2): 21–25.

SHAVELSON, R. J., AND G. P. BAXTER. 1992. "What We've Learned About Assessing Hands-On Science." *Educational Leadership* 5: 20–25.

8 Learning Cycles for Elementary School Science

Science content is usually divided into such disciplines as earth science, biology, chemistry, physics, astronomy, or geology. These disciplines represent organizing curricula. The learning cycles in this chapter are drawn from three of these disciplines: biological sciences, earth sciences, and physical sciences. Each part contains learning cycles for each elementary school grade from one through six. Some of the materials in all of the learning cycles included here were taken from *The Learning Cycle and Elementary School Science Teaching* (Renner and Marek, 1988). All eighteen learning cycles reflect the language appropriate for the grade level for which it was written. Exploration is labeled *gathering data;* term introduction is labeled *getting the idea;* and concept application is labeled *applying the idea.* To assist the teacher, special instructions, explanations for some of the more open-ended questions, and the idea (concept) of each learning cycle are shown in brackets. Each learning cycle is followed by *teaching suggestions.*

The learning cycles are to be used as a basic guide for implementing this model in classroom science teaching. Teachers may modify these learning cycles to incorporate methods addressed in previous chapters of this textbook. These learning cycle guides may also be edited to reflect teachers' original ideas and skills, and to meet the needs of various students and classroom settings. For example, a teacher may incorporate technology and add more probing questions into these learning cycles for a group of advanced-skills students. The same teacher may modify the learning cycles in different ways for students who are inexperienced with technology or who have more difficulty with the subject matter. These learning cycles may also serve as a foundation for developing integrated or thematic units. The teacher may engage students in reading trade books, writing creative stories, or using mathematical skills related to the concepts within these learning cycles. The teacher may embed a greater or lesser degree of scaffolding into the learning cycles for a particular group of students, and use techniques that will promote more meaningful learning. For each learning cycle, teachers must implement the most appropriate authentic assessments to measure concept understanding and process skill development among their particular groups of students.

Thus, the learning cycles in this chapter can be used *as is* for some groups of students and classroom environments. However, the learning cycles are *best* used as a foundation upon which teachers add their own creative ideas—specific to their individual classroom situations—that will best promote learning among their students.

Because this book is limited to six learning cycles for each of the biological, earth, and physical sciences, we were careful to develop a "sample" that would accurately represent each scientific discipline. To do this, we considered several questions when selecting the concepts for the learning cycles in this chapter: What concepts should be included in the biological, earth, and physical sciences? Are the concepts, principles, and topics of the learning cycles fundamental and central to the discipline? Can the investigations be conducted in elementary school classrooms?

Below are the concepts we have found usable in and appropriate for elementary school science. They are listed for the biological, earth, and physical sciences for grades one through six. Concepts for the learning cycles in Parts A, B, and C were selected from these lists. The language of each concept does not necessarily reflect the grade level for which it is intended.

Concepts from the Biological Sciences

Grade One

1. An aquarium is a place for living things to grow.
2. Living objects in an aquarium are called organisms.
3. Seeds come from plants.
4. Seeds grow into plants.
5. Plants have roots.
6. Many seeds are edible, for example, pecans, peanuts, green beans.
7. All animals need food, water, and a place to live.
8. Animals have properties.
9. Properties tell us about the kinds of animals.
10. People have properties that make them different.
11. People use their senses to find out about objects.

Grade Two

1. Seeds grow into plants, and plants grow and change; that sequence is called the plant's life cycle.
2. Every animal goes through a life cycle.
3. Some plants change with the time of year.

4. Plants grown in the light are different from plants grown in the dark.
5. Animals change as they grow older.
6. An animal's home is its *habitat.*
7. Plants have roots that help hold them in the earth and grow.
8. Living things in the aquarium grow and change.
9. People change as they grow older.
10. Babies grow teeth and these teeth change as people grow.

Grade Three

1. If a seed has light, water, and warmth, it will *germinate,* or begin to grow.
2. A plant gets minerals from the soil.
3. Plants use light and water to make food.
4. Seeds are often moved in many different ways to a place where they can grow.
5. Organisms of the same kind that live in the same place are known as a *population.*
6. Animals cannot make their own food so they eat plants or other animals.
7. Organisms eat one another. This is a sequence called a *food chain.* For example, a rabbit eats grass and a fox eats the rabbit.
8. A *food web* is formed when several food chains or parts of food chains are combined.
9. The bones in your body form a system called a *skeleton,* or *skeletal system,* which serves as a framework for your body.
10. The bones in your body are fastened together at places called *joints.*
11. Bones have special properties, such as size and shape, which determine their special uses.
12. The human backbone is known as the *spine,* and it is made up of small separate bones called *vertebrae.*
13. The muscles in your body make up a part of your body known as the *muscular system,* which helps move your body.

Grade Four

1. The properties of individuals within a population have great variation.

2. Organisms in the same population can have many variations; for example, there is no other person in the world exactly like you.
3. When breathing takes place, air enters special structures in the chest cavity called the *lungs.*
4. A person's activities influence breathing rates.
5. Blood is pumped through the body in regular beats called a *pulse,* which varies with age.
6. The dark spot in the eye, the *pupil,* is affected by the amount of light present.
7. The surroundings in which organisms live are called the *environment.*
8. Temperature, chemicals, and the amount of water present influence the environment and are called *environmental factors.*
9. People are able to live in varied environments in the world by controlling some factors of the environment.
10. *Adaptations* are special properties that help an animal survive in certain environments.
11. Animals can be grouped according to the kind of food they eat.
12. Foods are placed together because of the *food elements* they contain.
13. To be well nourished you should eat food from each food group every day.
14. Foods selected each day from each group provide a balanced diet.

Grade Five

1. When exact numbers of particular objects or traits are not needed or available, it is useful to make an estimate.
2. All organisms living in an area must have a source of food.
3. Mold is a fungus and may be made up of tiny organisms.
4. Certain organisms such as mold use organic material for food. These organisms cause organic material to *decay* or *decompose,* so they are called *decomposers.*
5. Yeast, like mold, feeds on organic material, and like mold and bacteria is a decomposer.
6. Food producers, food consumers, and decomposers make up a food cycle.
7. Organisms in an area interacting in this kind of food cycle relationship are called a *community.*

8. The process of converting food into a state the body can use is called *digestion.* The many organs that carry out this process are called the *digestive system.*

9. Food gives off energy when it is used by the body. This energy is measured in a unit called a *calorie.*

10. Calories of energy can be stored by the body.

11. Any substance essential to body functioning can be called a *nutrient.*

12. Your diet should be balanced between energy foods and foods your body needs for growing and for repairing damaged parts.

Grade Six

1. Seeing with two eyes is called *binocular vision.*

2. The hole in the eye that lets the light in is called the *pupil.*

3. Images focus on the *retina,* which is on the back of the eye.

4. The eye has a changeable convex lens that focuses images on the retina.

5. Your heart has valves and pushes blood through your body.

6. The rate of the heart beat is affected by body activity.

7. Breathing is controlled by the changeable volume inside the chest cavity.

8. The populations of organisms within an area interact with one another and are called a *biotic community.*

9. A biotic community interacting with environmental factors is an *ecosystem.*

10. Humans and other animals add carbon dioxide to the air.

11. The movement and use of gases in the environment are sometimes referred to as the oxygen-carbon dioxide cycle.

12. Any substance that in quantity has a bad effect on any organism can be called a *pollutant.*

Concepts from the Earth Sciences

Grade One

1. The sun gives light and heat, which warm the earth.

2. The moon is an object like the earth, but it is different, too.

3. Air is an object and has properties.

4. Clouds are objects.
5. Thunder is sound caused by lightning.
6. A rainbow is made of light of many colors.
7. Water is a liquid.
8. Water can be a solid called *ice*.
9. Rain and snow are made of water and come from clouds.
10. The land of the earth is made of rocks and soil.

Grade Two

1. The sun appears each morning, and this is called *sunrise*.
2. The sun disappears each evening, and this is called *sunset*.
3. Objects can stop some light, making a darker place called a *shadow*.
4. Light from the sun bounces off objects, and this is called *reflected light*.
5. The moon changes in shape, color, and position in the sky.
6. The sun and moon do the same things over and over again. Things repeated are called *cycles*.
7. Moving air is called *wind*.
8. The wind is described by how fast it is blowing and from which direction it is moving.
9. Clouds are different sized and shaped objects that move in the sky.
10. Temperature tells how hot or cold an object is. Some places on earth are cold, some are hot, and some are in between.
11. A year has twelve months. Each month has about thirty days, and the months make four groups called seasons.
12. The earth has a layer of air around it and a layer of water on it.

Grade Three

1. *Evaporation* is changing from a liquid to a gas; water can evaporate.
2. *Condensing* is changing from a gas to a liquid; water vapor can condense.
3. The water vapor in a cloud that condenses and falls to the earth is called *precipitation*.
4. How well things can be seen through the air is called *visibility*.

5. The air around us is known as the *atmosphere,* and it can have many conditions such as wind, moisture, and temperature.
6. Weather is the result of the conditions of the atmosphere.
7. The kind of weather in a place over a long period of time is known as *climate.* Climate is the average weather in a place, and there are different kinds of climate.
8. *Weathering* is the action of wind and water on the earth's surface and on its landforms.
9. When a forest fire destroys the trees, there is an open space on the land.
10. Cooled molten rock builds up into a mountain called a *volcano.*
11. When the ground at the surface shakes and trembles an *earthquake* is occurring.
12. Each of the four seasons during a year—spring, summer, fall, and winter—has its own kind of weather.
13. Plants and animals of long ago left their imprints, which are called *fossils,* in the rocks.
14. Dinosaurs, the largest land animals to live on earth, have become extinct, but we know a great deal about them from fossil evidence.

Grade Four

1. Water evaporates, and that vapor mixes with air, causing humidity.
2. Air can hold only a certain amount of water vapor at one temperature. If the air temperature gets higher, it can hold more, and if the temperature gets lower, it can hold less.
3. Atmospheric pressure is measured with an instrument called a *barometer.*
4. The amount of precipitation can be measured in inches or centimeters with a *rain gauge.*
5. Moving air is wind, and its direction can be measured with a *weather vane.*
6. Temperature, humidity, wind direction and speed, visibility, clouds, and precipitation are used to describe the conditions of the air that are called weather.
7. Weather conditions over a long time are called the climate.
8. Moving water and wind interact with objects such as rocks or soil and cause *erosion.*
9. Plants and animals decompose to produce *humus,* which, when mixed with sand, clay, and other minerals, makes soil.

10. Streams and rivers make up the drainage system of land.

11. A *lake* is an inland body of standing water in a depression or basin in the earth.

12. An *ocean* is a continuous body of salty water; oceans cover most of the surface of the earth.

13. *Rocks* are the materials that make up the earth's crust and form the mountains.

14. *Minerals* are the materials of which rocks are made.

15. Traces of animal and plant life preserved in rocks are called *fossils.*

Grade Five

1. The direction in which the numeral twelve on a clock is pointed when used to find objects is called *reference direction.*

2. A *compass* can be used to describe the location of an object.

3. The two numbers that tell where two lines cross are called the *rectangular coordinates* of that point.

4. An object that can be used to identify a particular place is called a *landmark.*

5. A *map* is a kind of drawing that shows the location of objects and places on earth.

6. Rectangular coordinates can be used with maps.

7. The position of an object or a place can be described as a certain number of degrees from a reference direction called a *polar coordinate.*

8. The location of an object or a place can be found if polar coordinates from two different positions are known. This is called *triangulation.*

9. A solid piece of matter near the surface of the earth can be called a rock and is made up of one or more pure substances called *minerals.*

10. The different characteristics of the earth's surface are called *surface features.*

11. The general shape of the land is called a *landform.*

12. The materials supplied by the earth are called *natural resources.*

13. Natural resources that can be used again and again are *renewable resources.*

14. The most abundant fossil fuel in the United States is coal, and enough coal reserves exist to last hundreds of years.

15. Saving and protecting our natural resources is known as *conservation.*

Grade Six

1. The properties of air that are especially interesting to weather scientists are called *weather elements* or elements of weather.
2. An *air mass* on earth is a huge body of air that can be hundreds of miles wide and many miles high and has the same temperature and humidity throughout.
3. An air mass with a lower temperature than the surrounding air is called a *cold* air mass, and an air mass with higher temperatures than the surrounding air is called a *warm* air mass.
4. When an air mass of a certain temperature moves into another air mass with a different temperature, the boundary between the two is called a *front.*
5. The model of the earth that best explains what people observe is a *sphere.*
6. The *sun cycle* changes position in the sky with respect to the earth from one part of the year to another.
7. There are four periods of the year called *seasons,* which can be differentiated from the earth-sun system model.
8. The sun is a star and gives light and heat to the earth.
9. A large spherical mass called the *moon* revolves around the earth.
10. An object that revolves in an orbit around a larger object is called a *satellite.*
11. Some objects in the sky are called *planets* and are satellites of the sun.

Concepts from the Physical Sciences

Grade One

1. All things are *objects.*
2. Color, shape, size, feel, and smell are *properties* of objects.
3. Objects are made of *material.*
4. Material is *matter.*
5. All matter takes up *space.*
6. Objects have their own sounds.
7. Sound moves in the air.
8. Moving air makes sound.
9. Music sounds have a pattern, while noise sounds have no pattern.

10. The clock tells the time things happen.

11. A minute and a second measure time. A minute is longer than a second and there are sixty minutes in an hour.

12. Machines make doing work easy.

13. The lever and the wheel are machines.

Grade Two

1. *Temperature* tells us how hot or cold an object is.

2. A *thermometer* may be a glass object containing liquid; heat makes the liquid move.

3. Pushing and pulling are *forces.*

4. Water digging up dirt and stones is *erosion.*

5. *Windstorms* are air moving with a lot of force.

6. All meter sticks have a balance point in the middle of the stick called the *center of balance.*

7. The longer the string of a pendulum, the slower the pendulum swings.

8. *Magnets* are special objects that stick to or attract iron.

9. One end of a suspended magnet points north.

10. A *compass* is a magnet.

11. Vibrating objects make sounds that travel through the air.

12. Strings vibrate to make sound.

Grade Three

1. The three states of matter are solid, liquid, and gas.

2. An electrical circuit provides a pathway for electricity, which lights a lightbulb.

3. The evidence of interaction among the elements in an electrical circuit can be the brightness of a lightbulb.

4. When electrical objects are put together and the bulb does not light, an open circuit is present.

5. Material that closes an open circuit is an *electrical conductor.*

6. When a magnet and an object pull toward each other, this pulling is called *magnetic attraction.*

7. Iron filings are tightly held to a magnet's poles.
8. Alike poles on magnets repel and opposite poles attract.
9. Electricity can cause magnetism.
10. Adding heat energy can make objects get hot and taking heat away lets objects get cool.
11. Heat can make other things happen in a system.
12. When talking about how hot or cold an object is, you are talking about its *temperature.*

Grade Four

1. What is seen in a mirror is called a *mirror image.*
2. A *line of symmetry* divides a shape into two *like parts.*
3. Different kinds of matter having different properties are called *substances.*
4. Matter that has variable properties is called a *mixture.*
5. If water and another material form a clear mixture when put together, the mixture is called a *solution.*
6. When a solvent has dissolved all of the solute it can hold, the mixture is called a *saturated solution.*
7. The transfer of heat energy through a material such as a metal rod is called *conduction,* and the material that allows the energy to be transferred through it is called a *conductor.*
8. Heat travels through air by *convection.*
9. There is an advantage to using the machine called a *lever* because the lever multiplies force.
10. A *ramp,* or *inclined plane,* multiplies force and is called a *machine.*

Grade Five

1. Energy is something that causes objects to do things they would not do without it.
2. One object or system is evidence of *energy transfer.* The object or system giving the energy is the energy *source,* and the object or system receiving the energy is an energy *receiver.*
3. Heat is a kind of energy.

4. Stored energy is *potential* energy.
5. Rubbing an object gives it an electrical charge, which is called *static electricity.*
6. Energy from dry cells is moving electrical energy and is called an *electric current.*
7. Energy produced when chemicals interact is called *chemical energy.*
8. Energy can cause things to vibrate, which produces sound.
9. When liquids of different temperatures are put together the resulting temperature—*equilibrium temperature*—is higher than the cooler one.
10. Machines can multiply force.
11. Energy can be changed from one form to another.
12. *Motion* is a change of position.
13. The push given to an object is called *action,* and the push the object gives back is called *reaction.*
14. Gravity attracts objects, often causing them to fall down.
15. The movement of an object moving horizontally and falling toward the earth at the same time is called *projectile motion.*

Grade Six

1. The smallest particle of a pure substance is called a *molecule.*
2. When an object increases in volume or size, the object *expands,* and when an object gets smaller or decreases in volume, the object *contracts.*
3. When an image is clear and sharp, it is said to be *in focus.*
4. When an object is far away, the distance from the lens to its image is the *focal length* of the lens.
5. The angle of a light ray going into a mirror is the same as the angle of the light ray reflected from the mirror.
6. The bending of light by glass (or any material) is called *refraction.*
7. One kind of matter moving through another kind of matter is known as *diffusion.*
8. Electricity is a form of energy because it can cause something to be done.
9. A wire, a dry cell, and a lightbulb can be arranged in a system that causes an *interaction*—the lightbulb lights. Such an arrangement is called an *electrical circuit.*

10. The interaction of the wire, the dry cell, and the lightbulb causes electrical current in the circuit.

11. Objects hooked together in a circuit in a way that allows the same amount of current to flow through each of them form a *series circuit.*

12. Objects in a circuit can be connected so that the energy source—the battery—affects them the same, but the current in each is different. Such a circuit is a *parallel circuit.*

13. The accuracy of a model depends upon the quality and quantity of information given the person building the model.

Learning Cycles for the Biological Sciences

Six concepts from those listed at the beginning of this chapter have been selected for the term introduction phases of the learning cycles that follow. These concepts are the following:

Grade 1: An *aquarium* is a place for living things to grow (biology concept #1).

Grade 2: Seeds grow into plants, and plants grow and change; that sequence is called the plant's *life cycle* (biology concept #1).

Grade 3: Organisms eat one another. This is a sequence called a *food chain.* For example, a rabbit eats grass and a fox eats the rabbit (biology concept #7).

Grade 4: When breathing takes place, air enters special structures in the chest cavity called the *lungs* (biology concept #3).

Grade 5: Certain organisms such as mold use organic material for food. These organisms cause organic material to *decay* or *decompose,* so they are called *decomposers* (biology concept #4).

Grade 6: Your heart has valves and pushes blood through your body (biology concept #5).

Learning Cycle for First Grade: Building an Aquarium

Today you will work with some living objects. You will make a place for the objects to live.

Gathering Data

Here are pictures of some of the objects. Name the objects.

Take a jar. You will need some sand. Wash the sand. Put the sand in the jar. Now put water in the jar.

What items did you place in the jar? What living things can be placed in the jar with the sand and water?

Fish can live in a jar of water. Plants and snails live there too. You can watch the living things in the jar. You can see what happens.

Put your living things in the jar. Put in the water plants. Plant them in the sand. Put fish in the jar. Put snails in the jar, too. Take care of your living things in the jar. Watch the fish. Watch the snails and plants. Tell what happens.

Getting the Idea

Name some things in the jar. How did you care for the things in the jar? What did you make when you put sand, water, and living things in the jar? [The idea: *An aquarium is a place for living things to grow.*]

Applying the Idea

You made an aquarium in a jar. You watched plants and animals live and grow in your aquarium. Now build a large aquarium in your classroom with your teacher. What will you need to put in your aquarium first? What animals and plants will you put in your aquarium?

Watch the animals and plants in your class aquarium. What happens to the plants? What happens to the animals? Draw pictures of your aquarium over time. What places in the world around you are like your class aquarium?

Visit a nearby pond or lake with your teacher. What plants and animals do you see in the pond or lake?

Checking Up

1. Name the things in the jar with sand and water.
2. Name the things in your class aquarium.
3. Name some things you have seen in a pond or lake.

Teaching Suggestions Here is a list of the materials you will need to set up the aquariums: aged water, fish food, fish net, guppies, sand, snails, water plants, and half-gallon jars. There should be an aquarium for every three or four children. Be sure to "age" the water by letting it stand in an open container for twenty-four hours. The aging allows harmful gases to escape from the water and permits the temperature of the water to approach that of the room.

Among the water plants that are suitable for an aquarium are the vallisneria, elodea, sagittaria, and cabomba. You can obtain these plants from a pet store.

Put the fish in the aquarium after the plants. Guppies are excellent aquarium fish, but any other fish that are easily maintained may be substituted. Put from two to four guppies in each jar, both females and males. Male guppies are usually smaller and more colorful than the females. Add several snails to the aquarium. Permit the children to put the aquariums in several different places around the room. If the aquariums are in different places, they will develop in different ways. (The amount of sunlight is one condition that makes a difference.)

It is important that each group records the items that were put in their aquarium and notes the date each item was added. A picture record is useful if the children are careful to show the exact number of organisms in the aquarium.

Fish in the aquarium may die. If this happens, *do not remove the dead organism.* Allow the children to observe and discuss the event. A snail may eat the dead fish. The children then can make inferences about the snail's sources of food. If a number of organisms die at the same time, this may be an indication of a badly polluted aquarium. Cloudiness of water usually indicates a concentration of bacteria resulting from the decay of excess food or dead organisms.

In building an aquarium and observing the interactions within this environment, the children will be developing concepts related to living organisms and their environments. Their learning will depend to a large extent upon your willingness to provide time for observation, discussion, and simple experimentation.

A teacher may be tempted at times to skimp on materials and to avoid child participation when dealing with living things in the classroom. But the obvious interest of the children and the ongoing opportunity for concrete learning make the value of these experiences apparent.

Building a large class aquarium is similar to preparing the small aquariums. However, the large aquarium may be able to sustain some different organisms, such as tadpoles, snails, and perhaps turtles. More plant varieties may also be added. Consult your local pet store for some ideas.

The children can observe changes and interactions in the class aquarium throughout the semester or school year. If possible, take children to a nearby pond or lake. Children will see the similarities between their in-class aquariums and real-world aquatic habitats.

Learning Cycle for Second Grade: Growing Plants

A plant is a living thing. Let's find out about plants. Let's see how they live. Let's see how they grow. Let's see how they change.

Gathering Data

Plant some pea seeds. When does the plant come up? How long does it take? Plant some bean seeds. When does the plant come up? How long does it take?

Getting the Idea

You planted a pea seed. A pea plant grew. The plant has pods. What is inside the pea pods? You planted a bean seed. A bean plant grew. The bean plant has bean pods. What is inside the bean pods? What can you now do with what is inside the pea pods and the bean pods? [The idea: *Seeds grow into plants and plants grow and change; this sequence is called the plant's life cycle.*]

Applying the Idea

We plant a seed. It grows into a plant. The plant grows more seeds. The new seeds can grow more plants. A tomato plant grows and changes. It grows from little seeds. Look at some tomato seeds. What do the tomato seeds do when you plant them? Tomato seeds change into tomato plants. What do the tomato plants make?

Checking Up

1. How many days did it take before your seed grew into a plant?
2. What is inside a bean pod?
3. What do we call the changes in seeds and plants?

Teaching Suggestions This learning cycle leads the children through an exploration of change in the plants they have been growing. It introduces them to the concept of the *life cycle.* It also helps them to realize that certain seeds produce certain plants; that is, a tomato plant does not grow from an orange seed. The students should grow other types of seeds in the concept-application portion of the learning cycle. If possible, they could dry their bean seeds from the pods and grow their dried seeds.

The following is a list of materials needed for this learning cycle: bean and pea seeds, potting soil, planting containers, spoons or small hand trowels, pea and bean pods, paper towels, support sticks for plants, and "twist'ems" or other fasteners for plants.

Plan regular observations of the plants. Ask the children to draw the plants as they develop. Save the drawings so you can use them later to help children recall the changes that took place during the plants' life cycle. You may wish to provide sticks to use as supports when the plants get tall enough. Twist'ems work well for fastening plants to support sticks.

After the plants mature, flowers will appear. From the flowers, pods will develop. Allow the children to open pods from the pea and bean plants on paper towels. By comparing seeds, the children can state that some are called pea pods because they contain pea seeds, and some are called bean pods for similar reasons. Ask them what they think the pea seeds and bean seeds will grow into if planted. Dry and save pea and bean seeds to plant later in the year, or the following year, to help students form relationships between the seeds they see in the pods and the actual plants that grow from these seeds. If you use the previous year's class seeds, you will explain (or ask) students what had to be done to prepare seeds for planting (e.g., dry them).

Be sure that students water the plants regularly and keep the plants near a light source. These activities will help them learn how to care for plants and keep them healthy.

Learning Cycle for Third Grade: Getting Food

Do you have a pet? If so, you must feed your pet. You must give it food. You must give it the food it needs. Animals in their natural habitats must find their own food. There is no one around to feed them. What food do animals eat? How do animals find their food?

Gathering Data

Fill a two-liter jar with clear water. Let the water stand for two days. Then put six guppies in the jar. Do not feed the guppies for three days.

Next, fill a small jar with water. Take one guppy from the two-liter jar and put it in the small jar. Then put several daphnia (water fleas) into the small jar. What happens?

Getting the Idea

What kind of living things are algae? What kind of living things are daphnia? What kind of living things are guppies? What did the daphnia eat? What did the guppies eat? What has happened can be shown in this way:

Algae → daphnia → guppies.

The arrow means "eaten by." Now read these diagrams:

1. Grass → rabbit → fox.
2. Corn → cow → people.
3. Seeds → sparrow → cat.
4. Leaves → caterpillar → robin.
5. Acorn → squirrel → coyote.

[The idea: *Organisms eat one another. This is a sequence called a food chain. For example, a rabbit eats grass and a fox eats the rabbit.*]

Applying the Idea

Daphnia eat algae, which are plants. Daphnia are animals. Some animals eat plants. Guppies eat daphnia. Guppies are animals, too. Some animals eat other animals.

Algae and daphnia make up populations. Guppies are another population. As you have observed, one population eats another population. Which population eats which? Food chains tell you.

Complete each of these food chains. Replace the question marks with the names of organisms. Make a record of each in your science notebook.

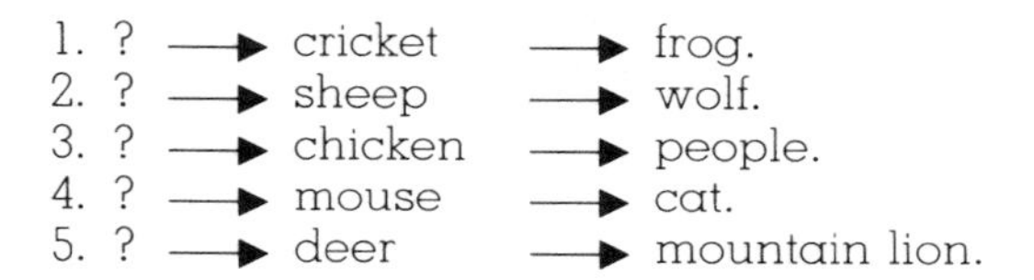

What goes in the place of each question mark? After you decide, write it down. How are the words you used for the question marks alike? The words you used for the question marks tell you something. What do they tell you?

The beginning of a food chain is always green plants. Green plants make their own food. They do not have to eat other organisms. Green plants combine water and carbon dioxide to make food. They do this with the help of light.

Each food chain has at least two animals in it. (A long food chain may have more than two animals.) One animal is eaten. The other animal does the eating. An animal that eats another live animal can be called a *predator*. The animal that is eaten is the *prey*.

Read this diagram:

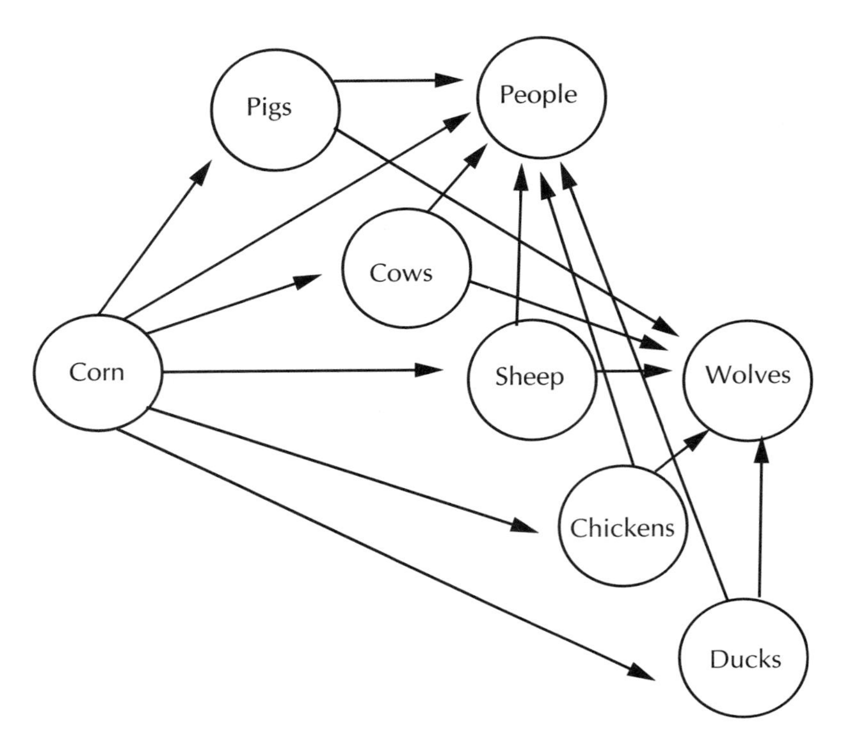

Which are the predators? Which are the prey?
Look at the list of organisms in the chart.

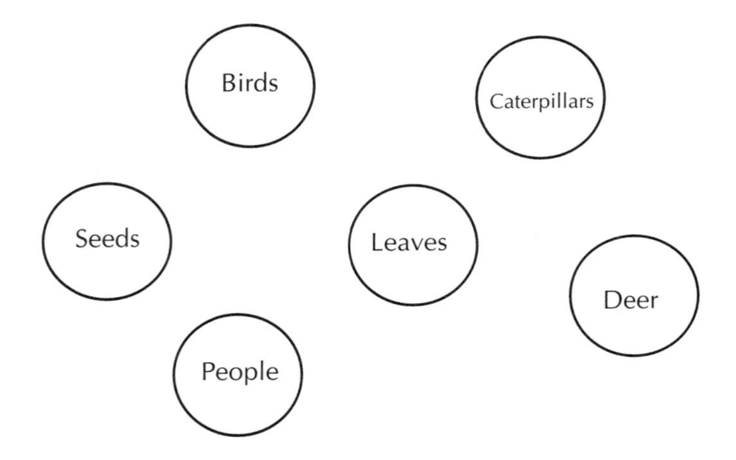

Make a food chain diagram. Which animal is the predator and which is the prey?

Checking Up

1. A daphnia is part of a food chain. What other organisms are you likely to find in the same food chain?
2. What kind of organism is always at one end of a food chain?
3. Which of the animals listed below are predators?

frog	cat
mouse	owl
sheep	wolf
chicken	cricket

4. Make a food chain for these three organisms: eagle, grass, gopher.
5. From where does a green plant get its food?

Teaching Suggestions The following is a list of materials needed for this learning cycle: daphnia, guppies, jars, paper towels, and eye droppers.

You can obtain daphnia from a supply house, or you can find some daphnia in a pond. Daphnia are commonly known as "water fleas," and can be found in great numbers around ponds and streams. Once you have a few daphnia, you can easily culture a large colony.

Instruct each group to put only one guppy into each jar with the daphnia. The guppies may be hungry enough to eat all the daphnia within two or three days.

The children have observed a food chain. (Do not use the name yet.) Daphnia are eaten by guppies. Introduce the arrow. Explain to the class that the arrow is a symbol that means "is eaten by." The diagrams represent three food chains. Algae are eaten by daphnia, and daphnia are eaten by guppies. Grass is eaten by a rabbit, and the rabbit is eaten by a fox. Corn is eaten by a cow, and the cow is eaten by people. Call to your students' attention that each food chain begins with a plant.

The children will carry out experiments that lead to the invention of a food web by interpreting diagrams. The children should think this exercise through carefully, and each child should be allowed to pick out predators and prey. Allow each child to take an animal through a particular sequence and find out when it is a predator and when it is prey.

The untangling of the food webs to distinguish predator and prey is a revealing exercise. The children must do this in order to become aware that a particular organism does not always stay in the same food chain.

Learning Cycle for Fourth Grade: Lungs

As a human organism, you are made up of many systems. There are many systems in your body. Your eyes, heart, and lungs are all parts of body systems.

Gathering Data

Put your hand on your ribs. Take a deep breath or inhale. What happens to the size of your chest? Now exhale. What happens to the size of your chest when you exhale?

Build the model described by your teacher. Pull the rubber covering on the bottom of the bottle down. What happens to the balloon? What have you done to the size of the "cavity" or space within the bottle? Now push the rubber covering into the bottle. What happens to the balloon this time? What have you done to the size of the cavity within the bottle this time?

Count the number of times that you breathe in one minute (one inhalation and one exhalation count as one breath). Make a record of that number. Now, repeat the measurement. Find out how many times you breathe in five minutes. Make a record of that number. How many times will you record that you breathe each minute?

Use the records of everyone in the class to make a class histogram. Choose a title for the histogram. Ask your teacher to explain what *rate* means. What is the range of the rates? What is the rate for the greatest number of people in your class?

Run hard in place for several minutes. Stop running. Have a friend count the number of times you breathe in one minute. Use records of everyone in the class to make a histogram of breathing rates after exercise.

Getting the Idea

Why does the size of your chest become larger when you inhale? Why does the size of your chest become smaller when you exhale? Refer to your model. What does each part of your model represent in your body?

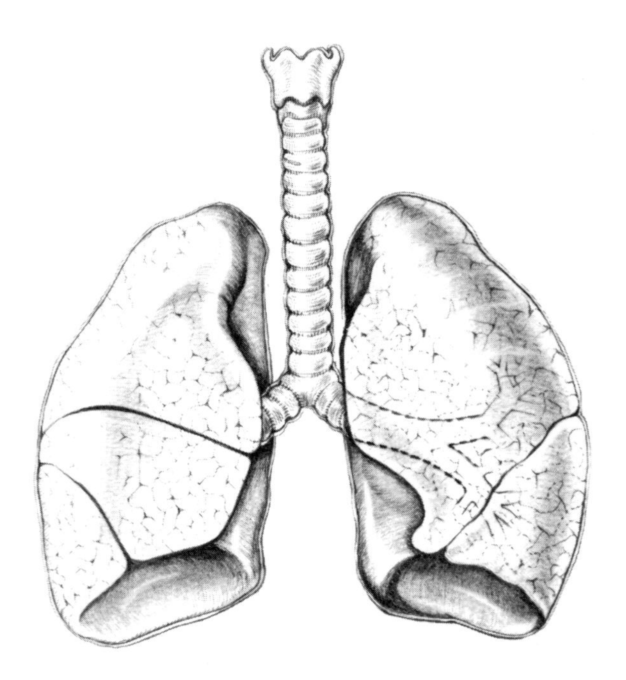

Compare the normal breathing rates of the class with the rates after exercise. Why are these two rates different? [When our chest cavity expands, air comes into the body. When our chest cavity contracts or gets smaller, air leaves the body. This process, called breathing, occurs faster when our bodies need more air, for example, when exercising.]

Where does the air "go" *within your chest cavity* when you inhale? Where does the air "leave from" when you exhale? [The idea: *When breathing takes place, air enters special structures in the chest cavity called the lungs.*]

Applying the Idea

Your lungs are expanding as they fill with air. Your lungs are made of a spongy tissue. This spongy tissue expands when you take in air. Then, when you breathe air out, your lungs squeeze together, or *contract*.

The human lungs are located in the chest cavity. One lung is on each side. The lungs' function is to furnish oxygen to the blood and to remove carbon dioxide.

Air enters through the nose and travels down the windpipe to the two branches of the lungs. The tubes leading to the lungs divide into smaller and smaller branches, ending in millions of tiny air sacs.

Tiny blood vessels surround these air sacs. It is here that the blood picks up oxygen and releases carbon dioxide. The carbon dioxide is breathed out of the body. The blood with a fresh supply of oxygen is returned to the heart to be pumped around the body.

Lung tissue is soft, light, and spongy. A thin membrane covers each lung. Breathing is done largely by a muscle called the diaphragm.

Children's lungs are usually pink, depending on the cleanliness of the air they breathe. The color of lung tissue varies with the environment. Some dust and dirt particles from the air remain trapped in the air sacs and cause a change in the appearance of lung tissue.

Lung tissue from people who smoke or who work in dusty places is found to be dark and less able to provide oxygen. These people are thought to be more likely to suffer from certain diseases.

When we breathe quietly, each breath contains about a pint of air. Great activity causes a change of this amount. We may take in as much as six times more air to provide enough oxygen.

Take as deep a breath as you can. Use that breath to blow up a balloon. Be sure you have blown as much as you can in one breath into the balloon. Tie the open end of the balloon securely. Tie it with a string.

Tie a nylon cord with two plastic rings as shown. Fishing line does nicely. Tape the rings to the balloon. Study the picture. Make sure the balloon moves freely along the cord.

Now, cut the balloon open. Cut off the end that is tied with a string. Use a pair of sharp scissors. What does the balloon do when you cut it open?

How does this experiment give you quantitative data? What do the quantitative data tell you?

Repeat the experiment until you find out the lung size of everyone in the class. Use your data to make a histogram. What is the variation in your classmates' lung sizes?

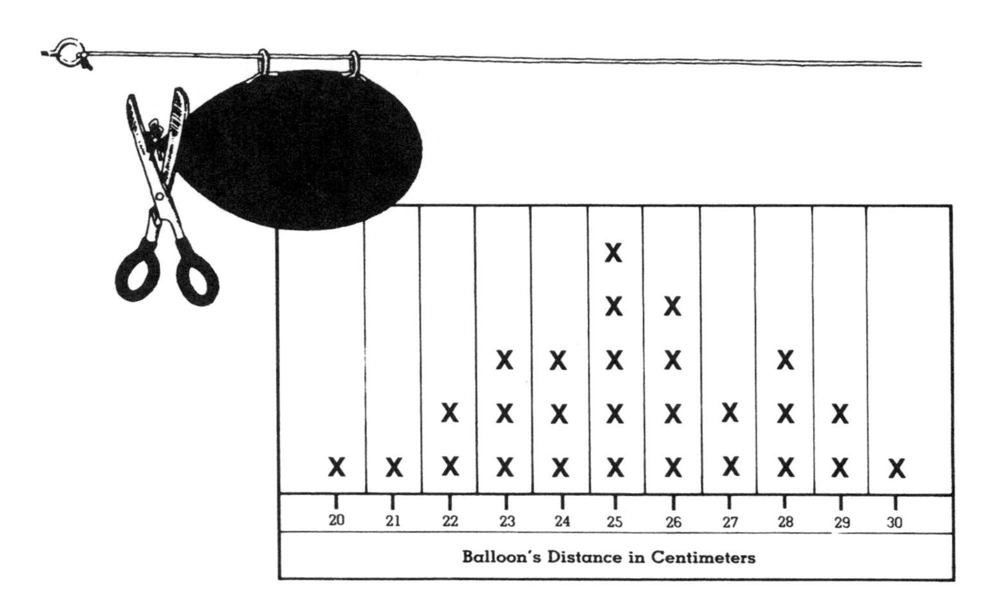

Lung size is often called *lung capacity*. Why is "capacity" a better word to use than "size"? Refer to a dictionary.

You now have a set of data about lung capacity. You can use your data to find out something else. First, measure the height of everyone in the class. Height and lung capacity are recorded in the table.

Look at the quantitative data you have found for the two variables. Compare everyone's height with their lung capacity. What do you see that might be important?

Student	Height	Lung Capacity

Checking Up

1. What happens to your chest when you take a deep breath?
2. What happens to your lungs when you breathe air in?
3. What happens to your lungs when you breathe air out?
4. Name something that might change your breathing rate.
5. Give some evidence about variation in lung capacity.

Teaching Suggestions The following is a list of materials necessary for this learning cycle: a one-liter clear plastic soda bottle with the bottom half cut off [a coffee can with the bottom cut out also works well, as shown in Chapter 1, Figure 1-5], a thin rubber sheet about fifteen centimeters square, a one-hole stopper, a glass Y-tube about nine centimeters long, balloons, and a large rubber band or masking tape.

The model is a simple apparatus. Two balloon (lungs) are sealed inside the plastic bottle (the chest cavity). The mouth of each balloon is fixed to the ends of the Y-tube (the trachea and bronchial tubes), which opens into the atmosphere, but not into the bottle. The single tube of the Y-tubing is pushed through the one-hole stopper. Some glycerin or mineral oil will help with the insertion of the tube into the hole of the stopper. The stopper should fit tightly into the opening of the bottle so that the only way air can enter the bottle will be through the end of the glass tubing. On the bottom of the bottle is a movable rubber covering (the diaphragm) affixed to the cut-off bottom with the rubber band or tape. When this covering is stretched, the interior volume of the bottle expands. When the rubber covering is pushed in, the bottle's volume is reduced.

Be sure your students understand these properties of the model:

1. The air can get into the inside of the balloon from the outside, but it cannot get into the inside of the bottle.
2. When the rubber piece on the bottom of the bottle is stretched, the area inside of the bottle becomes bigger.
3. When the rubber piece on the bottom of the bottle is pushed in, the area inside of the bottle gets smaller. (The "inside" of the bottle is the volume of the bottle.)

The ideas you need on pressure are very simple. If the air pressure on the inside of a container is lessened, air quickly moves in and raises the pressure back to normal. If the air pressure on the inside of a container is increased, air moves out and returns the pressure to normal. Be sure that the children understand that air always moves in order for pressure to remain as "normal" as possible (that is, a high-pressure area will tend to reduce its pressure and a low-pressure area will tend to have its pressure raised).

Students will develop a concept of breathing by interpreting the model. As the children will learn, air pressure and the movement of the chest make it possible for a person to breathe.

Learning Cycle for Fifth Grade: Decomposition

On what things is mold likely to grow? Will mold appear on leather? What about paper? Will a plastic cup grow mold? Does mold grow on wood? Will mold grow on butter, meat, and celery?

Gathering Data

You will need a mold terrarium to carry out this investigation. You can easily make the terrarium. Use any clear container with a tightly fitting lid. A gallon jar works well. Put in several inches of sand or dirt. Pour in enough water so the sand is damp. Choose the things you want to test. Then put them in the mold terrarium.

Keep a class record of all the materials placed in the mold terrarium. Be sure to record the date on which you placed each material in the terrarium. On which materials do you believe mold will grow? Make a prediction. Watch for the appearance of any growth and record the date. Keep the lid on the terrarium.

Your record of this experiment should look like this:

GROWTH IN MOLD TERRARIUM

Object	Date Put In	Prediction	Results	Date
Bread	March 4	Yes	Mold grew	March 9

Keep your mold terrarium growing for as long as you observe changes. What finally happens to the things inside the container? Check your predictions about the different materials.

Now it's time to open the terrariums. Dump the materials out and carefully observe the condition of the materials. Spread out the contents on a paper towel. Observe the materials with a hand lens. Compare the contents of your terrarium with the contents of other terrariums.

Getting the Idea

What happened to the amount of organic material—for example, bread—you put in the mold terrarium after a period of time? What happened to the amount of mold in your terrarium after time? Why was the amount of organic material different at the end of this experiment? [The organic material was *less* because the mold "ate" or digested the material. The mold uses organic material such as bread for food. The amount of mold was greater at the end of the experiment because it was able to grow and reproduce.] [The idea: *Certain organisms such as mold use organic material for food. These organisms cause the organic material to "decay" or decompose, so they are called decomposers.*]

Applying the Idea

To *compose* something means *to put together.* The prefix *de* means *the opposite of.* So the word *decompose* means *to take apart.* When mold grows on something it uses that material for food. It takes what it needs for its own growth. In doing so, the material is taken apart, or *decomposes.*

Decomposition is an ongoing process. Mold spores are present all around. If environmental factors are favorable, the spores grow and decompose whatever organic material they contact.

Molds grow in varied shapes and colors. Bread mold is usually seen as a black, cottony growth. It can obtain food from living or nonliving materials. It is one of the most active decomposers.

Some molds can be called *parasites.* A parasite is an organism that lives on living material. These parasite molds are the cause of much decomposition in fruit and vegetables. Plant molds destroy large amounts of farm crops.

Certain molds are present in soil and water. Water molds live on the bodies of fish and insects. Greenhouses and fish hatcheries are interested in the control of these fungi.

One of the most destructive fungi is the one that attacks potato plants. It grows on the leaves of the plant and can destroy an entire crop.

When decomposers are not controlled they can be harmful. Remember, however, their part in decomposing unwanted materials. Decomposition is a process necessary to the well-being of every environment.

Checking Up

1. Describe what happened in your model terrarium.
2. What materials molded first?
3. What does *decompose* mean?
4. In what way is mold a decomposer?
5. In your opinion, is mold a helpful or harmful organism? Explain your answer.

Teaching Suggestions The following is a list of materials needed for this learning cycle: a gallon jar with lid, sand or dirt, water, and materials to test (paper, plastic, foods, leather piece).

Students are usually familiar with terrariums as containers in which plants grow. Explain that they now will build a mold terrarium. Decide how many terrariums you would like to have in the classroom. One would be adequate, but, if your class is large, you may need two to accommodate all the things that the students will want to test. Be sure the container has a lid. Wide-mouthed gallon jars work well and are inexpensive.

Discuss what to put inside the terrarium for a base. Soil or sand is fine. The terrarium should be damp and it should be set in a warm place. Question the children about what materials should be placed inside. The students may suggest items on which they know mold will grow. They will test the items by putting them in the terrarium. Instruct the children to keep a record of the materials they test. Refer to the chart under Gathering Data in the learning cycle as a means of keeping records. Call attention to the Prediction column on the chart.

Students may wish to keep adding items to the mold terrarium. Allow them to do this only until a good growth of mold occurs. Once there is considerable mold in the jar, discourage your students from taking off the lid. Discuss the changes that occur.

As a final activity with mold, have the children open the terrarium and examine the contents. The organic material should now be in a state of decomposition. The observation leads you directly into the students' conceptual development of decomposition and into their perception of mold as a decomposer.

Learning Cycle for Sixth Grade: How the Heart Works

Put the finger of your left hand on the thumb side of your right wrist. You will feel a thump, thump, thump. You probably know that you have just felt your pulse. Your pulse, as you know, has something to do with blood and the heart.

You know more about your body than people in the year 150 A. D. knew. In those days everyone felt that the thump on the wrist had something to do with air. A physician named Galen then found out that blood causes the pulse, not air.

Galen wrote many books. He studied how the human body works. Galen had many good ideas, but he did get some things wrong. For one thing, he gave a wrong explanation of how blood moves through the body.

An English physician, William Harvey (1578–1657) provided a better explanation. Galen was a famous man, and many people went along with his ideas instead of Harvey's. Still, Harvey's explanation proved to be right.

For many years, Harvey had studied how blood moves around the body. In 1628, he wrote a book showing Galen to be wrong. Just imagine! More than 1,400 years had passed since Galen's time! Sometimes even when ideas are wrong, they last a long while.

But what did William Harvey find out? What were his ideas? What kind of model did he build for blood moving throughout bodies? Harvey first suggested the model we use today. There have been some small changes, but the model is basically like the one that Harvey proposed. Let's find out about Harvey's famous model.

Gathering Data

You need to build equipment to use in the following experiment. There are some special things you will need. Here is a list of what you will need for the experiment:

1. Get a plastic bottle like the one pictured below. Be sure the bottle is soft enough to squeeze.
2. You will need two rubber stoppers, each with one hole.
3. Get two pieces of glass tubing, each about eight centimeters long.

4. You will need two pieces of rubber or plastic tubing, each about twenty centimeters long.
5. Get two jars. Each one should hold about three hundred cubic centimeters.
6. You will need a third piece of rubber tubing about fifty centimeters long.

Now, put your equipment together. Follow these directions:

1. Hold a rubber stopper and a piece of glass tubing under water. Gently twist the glass tube through the hole in the stopper. Then do the same with the other rubber stopper and glass tube.
2. Cut a hole in the bottom of the plastic bottle. Fit one of the rubber stoppers into the hole.
3. Take the cover off the bottle. Fit the second rubber stopper into the hole in the top of the bottle.
4. Now, attach a piece of rubber tubing to each glass tube. Your equipment should look like this:

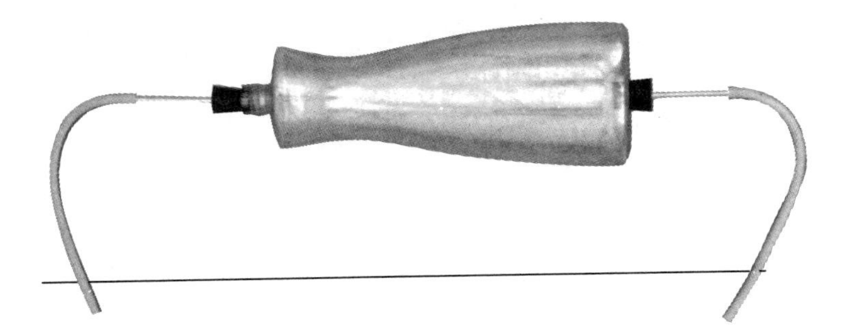

Next, half fill one of the jars with water. Call this *Jar 1*. Leave the other jar empty. Call this *Jar 2*. Put one of the pieces of tubing hanging from the bottle into Jar 1. Put the tubing on the other end of the bottle into Jar 2. Call the tube in Jar 1 Tube A. Call the other Tube B.

Now you are ready to use your model. You will need a partner.

Put water into the bottle until it is about half full. To do this, you can remove the stopper at the top of the bottle. Squeeze Tube B tightly so that the water does not leave the bottle while it is being filled. Hold the bottle on its side. Squeeze Tube A with your fingers.

Now, give the bottle a quick, hard squeeze. Do not release the bottle. Observe what happens. Hold the end of Tube B above Jar 2. Now, release the bottle. Repeat this until the water in the bottle is about half gone.

Tightly squeeze Tube B. Squeeze the bottle and release it. What happens to the water in Jar l? Repeat this several times. Make a careful record of what happens when each of the tubes is squeezed tightly.

You also have a piece of rubber tubing about fifty centimeters long. Put one end of that tubing into the jar Tube B has been emptying into. Suck the air out of the long tubing. Start the water running from Jar 2 to Jar 1. Practice as much as you need to. Keep the water running between the jars.

Now your equipment is working. Water is moving from Jar 1 to Jar 2 through the bottle. You know how to squeeze the bottle and the tubes to do this. Practice until you can keep the amounts of water in Jars 1 and 2 the same.

Describe how the water goes from Jar 1 to Jar 2. Compare how the water reaches Jar 2 with how it goes from Jar 2 to Jar 1. Be sure to compare the way the water flows in both cases. Make accurate records about the differences in the flow from Jar 1 to Jar 2 and from Jar 2 to Jar 1.

Getting the Idea

Again, feel your pulse. What in your experiment could be compared with the thump of your pulse? What caused the short burst of water that entered Jar 2? Use your data to answer these questions.

What in your body causes the thump of your pulse? Use the data you have gathered to make a model of how your heart pushes blood though your body. How does your model allow you to explain heartbeat?

Your fingers were an important aid in moving water from Jar 1 to Jar 2. Explain their purpose in the experiment. [The idea: *Your heart has valves and pushes blood through your body.*]

Applying the Idea

The valves in your heart keep blood in its proper part. Your fingers worked as valves in your experiment. Your heart delivers blood to the *arteries* in your body. It delivers blood to the arteries in spurts.

The heart is a pump. It is an organ that pumps blood through the body. As a pump, the heart is also a muscle. It is a muscle that keeps working automatically.

The human heart is made up of four separate parts. All mammals have hearts consisting of four parts. The heart has a right side and a left side. A wall known as the *septum* lies between the two sides.

The blood leaves the heart through the aorta. The *aorta* is the principal blood-distributing *artery* in the body. Smaller arteries branch from the aorta. They carry blood to the head, arms, legs, intestines, liver, kidneys, and stomach.

In all parts of the body, there are small blood vessels called *capillaries*. The blood flows from the arteries into the capillaries. From the capillaries the blood flows into the *veins*. The veins carry the blood back to the heart.

The blood flows from the veins into the right *atrium*, an upper part of the heart. The blood then begins its flow through the heart itself. From the right atrium it drops into the right ventricle. The *ventricle* is a lower part of the heart. The blood flows from the right atrium to the right ventricle through the *tricuspid* valve.

All the blood flows into the right ventricle, and the tricuspid valve closes. The blood then flows through the *semilunar valve* into the *pulmonary artery*. The *pulmonary artery* carries the blood to the lungs.

The lungs remove carbon dioxide from the blood. The blood has picked up the carbon dioxide as a waste product from the body. The carbon dioxide is exchanged for oxygen. With a fresh supply of oxygen for the body, the blood then flows from the lungs into the *pulmonary veins*.

The blood returns to the heart through the pulmonary veins. It flows from the pulmonary veins into the left atrium. The *mitral valve* opens. The blood then flows into the left ventricle.

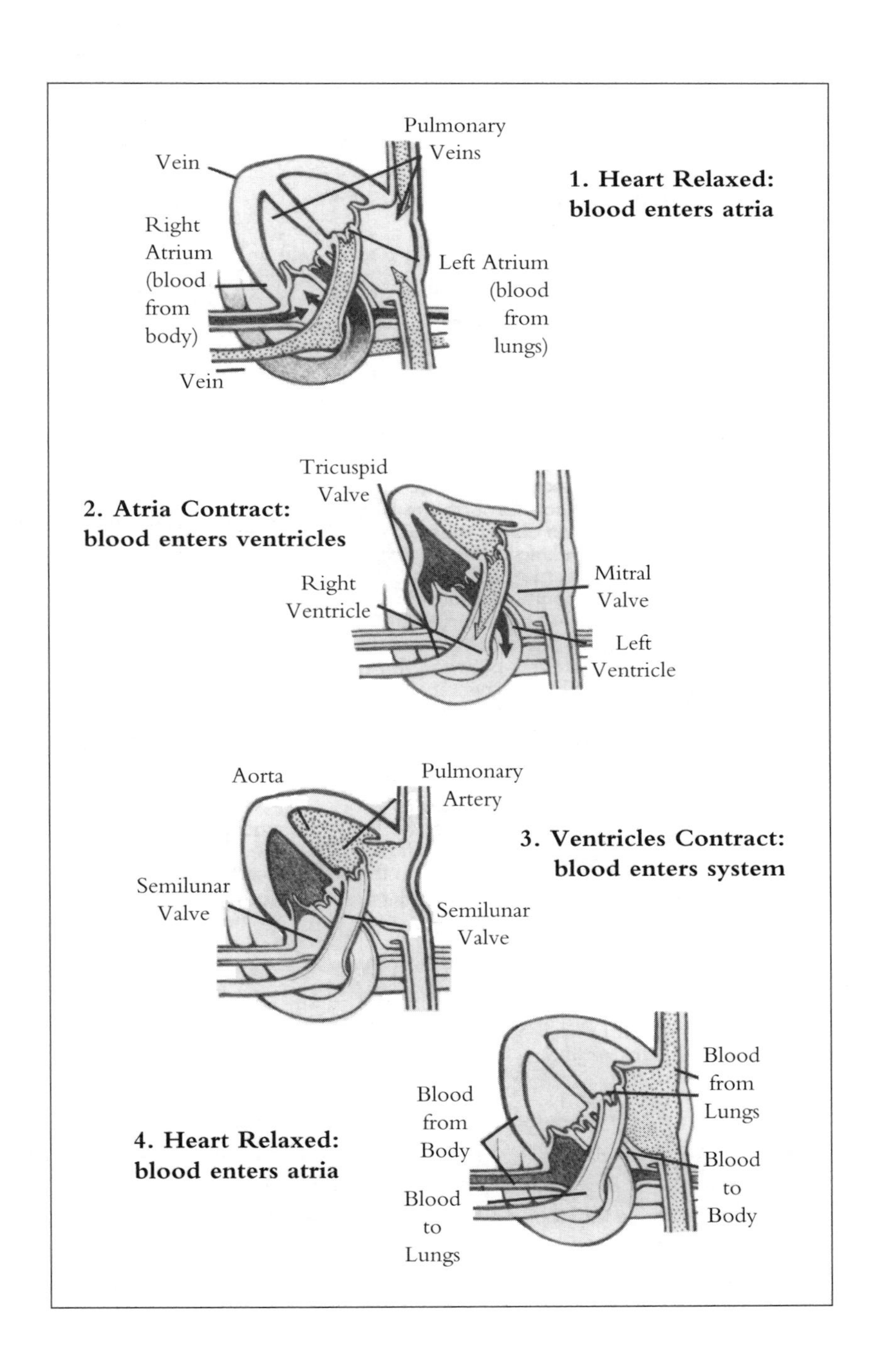
Pulmonary
Veins
Vein
1. Heart Relaxed:
blood enters atria
Right
Atrium
(blood
from
body)
Left Atrium
(blood
from
lungs)
Vein
Tricuspid
Valve
2. Atria Contract:
blood enters ventricles
Right
Ventricle
Mitral
Valve
Left
Ventricle
Aorta
Pulmonary
Artery
3. Ventricles Contract:
blood enters system
Semilunar
Valve
Semilunar
Valve
Blood
from
Lungs
Blood
from
Body
4. Heart Relaxed:
blood enters atria
Blood
to
Body
Blood
to
Lungs

The mitral valve closes. The left semilunar valve then opens, and the blood is pushed into the aorta. The process starts over. You must remember that both sides of the heart work at the same time. You have been led through the circulation here one stage at a time.

In the capillaries, the blood picks up waste products including carbon dioxide. The kidneys remove much waste from the blood. The lungs remove carbon dioxide and put oxygen into the blood.

Checking Up

1. What can you feel when you put your finger on the thumb side of your right wrist?
2. Where does the blood go when it leaves the heart?
3. To what part of the body do the veins deliver blood?
4. What purpose do the heart valves serve?
5. Explain the difference between veins and arteries.

Teaching Suggestions The following is a list of materials needed for this learning cycle: glass tubing (two pieces, eight centimeters long), two jars (three hundred cubic centimeter capacity), a plastic bottle, two rubber stoppers (one-hole), rubber tubing (or plastic tubing; two pieces, twenty centimeters long), and rubber tubing (one piece, fifty centimeters long).

The students in your class will construct a model of the circulatory system. With this model, they will do an experiment that shows how the heart functions. The students will learn about the valves of the heart and how the valves control the flow of blood. The activity is also an investigation of pulse rates.

The model is actually a siphon, which causes water to flow from one jar to another, just as the heart pumps blood to the body and back to the heart. The children must be sure that the water is started properly. To ensure a smooth flow, they must constantly compress and release the plastic bottle. This compression and release corresponds to the heartbeat. The water flowing through the bottle from one jar to the other is the "blood" and it moves in spurts.

It will probably take at least two class periods for the children to get their apparatus constructed and functioning properly. Children of this age may be tempted to shoot water at one another and to engage in water fights. Be aware of this possibility.

The children are inventing the concept of heart valves based on their concrete experience. Their fingers acted as valves when they removed them from one side of the plastic bottle and placed them on the other side in order to stop the flow of water in one place or to start it in another place.

Call the children's attention to the fact that they have constructed a model of the circulatory system and the heart. The tube through which the water flows from one jar to another represents the veins. Jar 1, Tube A, and the plastic bottle are the heart. Jar 2 is the body. Thus, we have "blood" flowing from the heart to the body and back to the heart.

Learning Cycles for the Earth Sciences

Six concepts from those listed at the beginning of this chapter have been selected for the term introduction phases of the learning cycles that follow. These concepts are the following:

Grade 1: The land of the earth is made of rocks and soil (earth science concept #10).

Grade 2: A year has twelve months. Each month has about thirty days, and the months make four groups, called *seasons* (earth science concept #11).

Grade 3: Cooled molten rock builds up into a mountain called a *volcano* (earth science concept #10).

Grade 4: Traces of animal and plant life preserved in rocks are called *fossils* (earth science concept #15).

Grade 5: A *map* is a kind of drawing that shows the location of objects and places on earth (earth science concept #5).

Grade 6: When an air mass of a certain temperature moves into another air mass with a different temperature, the boundary between the two is called a *front* (earth science concept #4).

Learning Cycle for First Grade: Soil and Rocks

Above the earth is air. There are clouds too. On the earth there is water. But the earth is made of something else.

Gathering Data

Look around you. What is the earth made of? What do you walk on? What holds trees? What do flowers and grass grow in? What are houses built on? What is at the bottom of a river? What is a mountain made of?

Find some rocks. Find rocks of different colors. Find rocks of different sizes. Look for a very small rock. Look for a very big rock. Tell where you saw them. Look at a rock. Use a hand lens. Tell what you see. Feel some rocks. Tell how they feel. Make a group of smooth rocks. Make a group of rough rocks.

Look at the soil your teacher (or students) brought in to class. Feel the soil and see what is in it. Where did the soil come from? What things are in the soil?

Look at some pictures of mountains. Look at some pictures of land. Tell what you see.

Getting the Idea

Where did you find the rocks? Where did you find the soil? What are mountains made of? What is land made of? What is at the bottom of a river? [The idea: *The land of the earth is made of rocks and soil.*]

Applying the Idea

Rocks are hard chunks of matter. Some rocks are very large. Mountains are mostly rock. Some rocks are very small. A tiny grain of sand is a rock.

There are different kinds of rock. There is hard rock like granite. There is soft rock like talc. There is light rock like pumice. But most rock is heavy.

Rocks are found in many colors. Some, such as quartz, are clear. Some, such as lava, are black. There are green, red, blue, and yellow rocks too.

Soil is called *dirt*. It is what plants live in. Soil also has many colors. Soil can be red or yellow. Soil can be black or orange. Soil has tiny rocks in it. They are

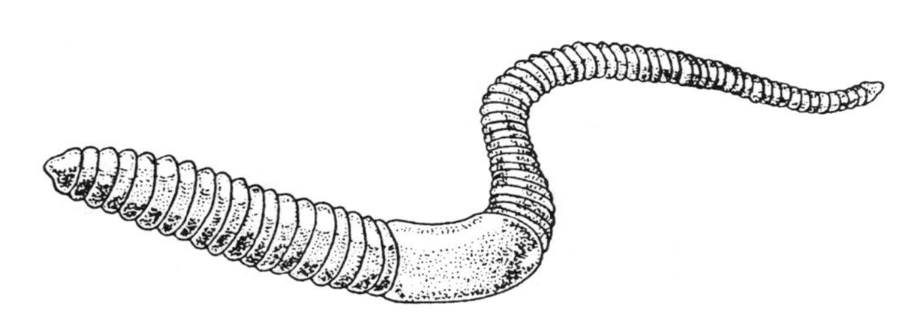

like grains of sand. Look at these pictures of animals. Some live in soil. Some live in rock caves. What other animals live in soil? What other animals live in rocks?

Farmers use soil. They plant crops. They grow wheat and corn. They grow vegetables. People use the crops for food.

Checking Up

1. Name two properties of rocks.
2. What colors are rocks?
3. Name an animal that lives in soil.
4. How do farmers use soil?
5. How are rocks different?

Teaching Suggestions The following is a list of materials needed for this learning cycle: hand lens, pan, rocks (variety of colors and textures), rock kits (optional) available from a supply house, and sand.

Much of the learning cycle can be conducted outside around the school or in a nearby field. The children can collect rocks and soil and make many of their observations. Their collections can be supplemented with pictures of rocks, fields, and landforms (mountains, canyons, hills, and volcanoes).

Some of the children may already have a variety of rocks they have collected. You may want to use their rocks or purchase rock kits, which are available through science supply organizations. Excellent rock and mineral samples are provided in these kits—granite, talc, pumice, and mica. It is not important that the children identify these rocks and minerals, but they should experience the properties of the samples—color, texture, and hardness, for example.

Learning Cycle for Second Grade: The Year

Some changes on earth happen every day. The sun rises and sets. The day gets light. Then it gets dark.

Some changes happen each month. The moon is full. Then it appears to change shapes. It becomes full again.

The daily change is a cycle. The monthly change is a cycle. The earth has another cycle too.

Gathering Data

Look at a calendar. Count the number of pages. Each page is one month. Look at each page. Count the days. How many days are on each page? How many days make one month? Do all months have the same number of days? Write the names of the months on cards. Make groups of the months. Group them by how many days they have. Each month has a name. We use the name to tell about things. Christmas comes in December. Valentine's Day comes in February. Thanksgiving is in November. What happens in other months? Tell something. What month were you born?

Put the cards into groups. Put January and February and March together. They make one group.

Put April and May and June together. Make a group of July, August, and September.

Make another group of October, November, and December.

Getting the Idea

What happens in the spring? What happens in the summer, fall, and winter? Look at the groups of months you have made. Which group do you think is spring? Which group is summer? Which group is fall? Which group is winter? [The idea: *A year has twelve months. Each month has about thirty days, and the months make four groups, called seasons.*]

January February March	April May June
WINTER	**SPRING**
July August September	October November December
SUMMER	**FALL**

Applying the Idea

Each year begins the first day in January. We call this New Year's Day. It is in the winter season.

It is usually cold in the winter. People wear coats to keep warm. The wind blows. Sometimes it snows.

Then comes spring. Puffy clouds fill the sky. Rains come. The temperature is warmer each day. People plant seeds. Soon the seeds will sprout and begin to grow.

The summer season is often hot. Sometimes there is not enough rain. But it is a fun time of year.

Fall is harvest time. Wheat and corn are ripe. The air begins to cool. Soon winter will come again.

The seasons happen over and over. They are a cycle. The whole cycle is one year.

Find out more about the seasons. What things happen in the spring? What things happen in the fall? What things happen in the summer and winter?

You name them. The teacher will make a list on the chalkboard.

Spring Things	Summer Things	Fall Things	Winter Things

Checking Up

1. How many months are in one year?
2. Name the four seasons.
3. What months are in the spring? Fall?
4. What happens in the spring? Fall? Winter?
5. How many days are in the longest months?

Teaching Suggestions The following is a list of materials needed for this learning cycle: calendars and four-by-six-inch index cards.

Time is a concept, but it is hard to define. Probably the best definition of time is *the space between any two events.* But you will immediately recognize great inadequacies in that definition. A second definition might be *the period during which something happens.* The Christmas season is an example.

Part of the trouble is that the definition of time must almost inevitably employ "space" or "period" or a similar word. In effect, we are defining time with examples using time. This process is illogical and improper. By now, you have probably concluded that there is no way in which time can be succinctly defined, and that all the definitions available are unsuitable for concrete operational thinkers. This leaves one alternative for you and your second-graders: they must experience time.

In this learning cycle, you will be using the students' experiences of the different events that occur in different seasons. Changes in temperature during the year are a good example. Ask the children what temperatures characterize the seasons. Have them correlate the seasons and the seasonal temperatures. You might like to show the connection graphically on the board (e.g., a picture of a girl and boy perspiring next to a blazing sun). Sports activities and holidays are also useful to develop and stress the correlation.

Learning Cycle for Third Grade: Volcanoes

Weathering and erosion wear down the earth. With constant weathering and erosion, landforms change. The earth changes. Other forces build up the earth. Sometimes these forces work slowly. At other times, these forces act quickly. What builds up the earth?

Gathering Data

Take a large piece of waxed paper. Make a spot in the center of the waxed paper with a marker.

Your teacher will give you some hot fudge in a clear, plastic measuring container. Pour the fudge directly on the marker spot.

Before the fudge completely cools, use a spoon to make a narrow hole in the fudge. The hole should be made directly over the original marker spot on the paper. Dig the hole until you can see the marker spot. Now let the fudge completely cool and dry.

Your teacher will give you some more hot fudge in your container. Pour the new hot fudge over the hole. Before the fudge completely cools and dries, dig the hole again. Now let the fudge completely cool and dry.

Your teacher will fill your container with the hot fudge again. Pour the hot fudge over the hole. This time, do not dig the hole. Let the fudge completely cool and dry.

Getting the Idea

What happened to the hot fudge as it cooled and dried on the waxed paper? What happened when you poured the new hot fudge over the hole in the dry fudge?

What material of the earth did the hot fudge look like? What was the shape of the "landform" that was building up after each layer of fudge (molten rock) was added? [The idea: *Cooled molten rock builds up into a mountain called a volcano.*]

Applying the Idea

The inside of the earth is hot. At a depth of fifty kilometers, the temperature is 700 degrees Celsius. The extreme heat melts rock deep below the surface of the earth. The hot, molten rock is called *magma*. Magma seeps into the underground cracks and fissures of the earth.

Sometimes a deep crack forms in the earth. Magma flows into the crack and pushes its way up to the surface of the earth. When the magma shoots out, it is called *lava*. The place where it comes out is a *volcano*. Lava spews from an erupting volcano.

Lava, smoke, steam, and ashes shoot into the air. As more and more lava streams out, a mountain begins to form. When the lava cools, it becomes solid rock again. In a short time, a volcano can grow to a height of many meters.

Volcanoes formed the land that is now the state of Hawaii. Nearly all the volcanoes in Hawaii have been inactive for many years. But one of them still *erupts*, or pours out lava, from time to time. The name of the active volcano is *Mauna Loa*. Mauna Loa builds up the land each time it erupts. It is more than 4,600 meters high.

Checking Up

1. What is the difference between magma and lava?
2. What happens when a volcano erupts?
3. How does a volcano change the land?

Teaching Suggestions The following is a list of materials needed for this learning cycle: recipe and ingredients for fudge mix, markers, waxed paper, spoons, and plastic containers (with spout for pouring, if possible).

Building a model of a volcano can be a messy investigation. You may want to collect an ample supply of newspaper to cover the floor. The children will be pouring fudge and must be cautioned not to touch the fudge while it is hot. Pot holders can be brought in by the children to use in this activity. The children can observe the teacher heating the fudge in the pan while they are waiting for their fudge to cool and dry. The teacher can ask students questions about their observations of bubbles and steam as related to molten rock and volcanic activity. Do not leave the fudge unattended, and caution students not to touch the pan. This activity would be a good one to invite a few parents or older students to attend. The parents or older students can help make the fudge and help monitor the students while you observe and direct the children. Afterwards, the children can eat the fudge.

The teacher could demonstrate a working volcano after the children have built their own volcanoes. Directions to construct a variety of different "working" volcanoes can be found in most earth science textbooks or laboratory manuals.

Most areas do not have volcanoes (either active or extinct), so the children may not be able to see a volcano. Mt. St. Helens is an active volcano in Washington, and an extinct volcano (Pilots Knob) exists in central Texas. Audiovisual materials (e.g., films or video) may be used in the concept application to show students an actual volcanic eruption.

Learning Cycle for Fourth Grade: Fossils

People learn about the earth from rocks. That is why geologists study them. They look for information about the earth's surface. They look for *evidence* of changes in the surface. They look for *evidence* of changes in the climate. And they look for *evidence* of things that lived long ago. Rocks tell the history of the earth.

Gathering Data

Look at this stack of playing cards. Suppose each card was laid down one at a time. Which card was laid down first? Which one was next? In what order were the rest laid down?

Look at these layers of rock. Which of these layers were laid down first? The study of layers of rock is called *stratigraphy*.

Some geologists study the rocks themselves. This is called *petrology*. Every rock can tell a story. Look at these pictures of rocks. What story does each tell?

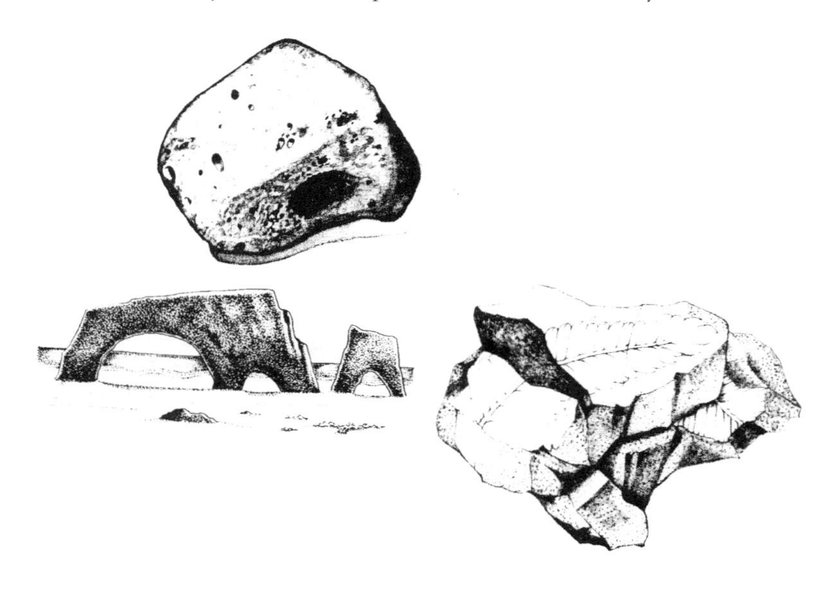

Work together in a group of three or four persons. Mix some plaster of paris with water. Make it like thick pancake batter. Make about one cup. When plaster of paris dries, it hardens like a rock. Pour the plaster of paris into a flat container. Smooth the surface. Then push some object into the plaster of paris. Make an imprint. Take the object out and let the plaster of paris dry. Let others try to figure out from the imprint what the object was. Let each person tell about the object from the imprint.

Geologists use rocks to tell what has happened on the earth. Layers of rocks tell the order in which things happened. The types of rocks tell about what kinds of things happened.

Getting the Idea

When you took your object out of the plaster, how could others tell what the object looked like? When animals and plants are within rock layers, how do you think scientists can tell what the animal or plant looked like? [The idea: *Traces of animal and plant life preserved in rocks are called fossils.*]

Applying the Idea

People can learn a great deal about the earth by studying sedimentary rock. Evidence of plants and animals called *fossils* are almost always found in sedimentary rock. Paleontologists study fossils to see what kinds of animals and plants lived while the rock was being formed. They can tell the size and shape of the animals. They can even tell the climate during a certain period of time. In some cases, geologists find evidence for an ancient ocean where land is now. A layer of sedimentary rock with sea shells in it is evidence. Rocks show that even the tops of some mountains were once under the sea.

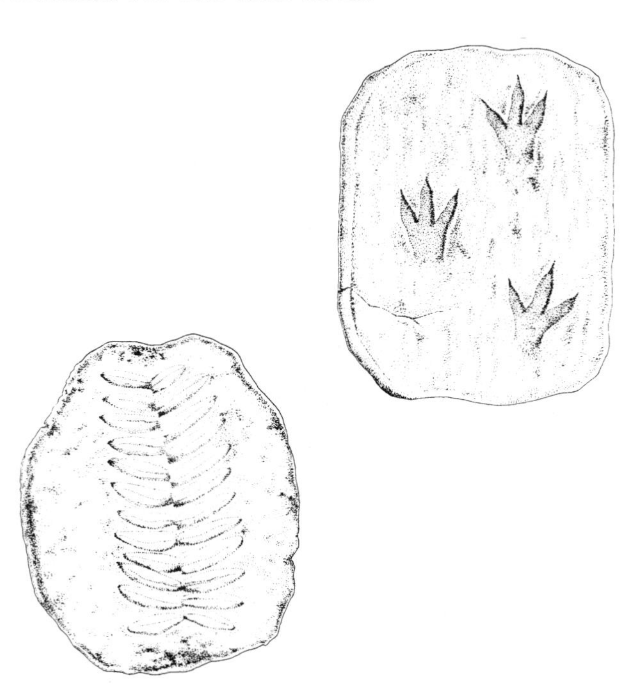

Go on a fossil hunt. Look in the same places you looked for rocks. Make a collection of fossils. Remember, anything that gives evidence of ancient life is called a fossil. A fossil can be just the tracks of an animal or the imprint of the organism.

A fossil can be a piece of skin or a shell; it can be a bone. A fossil can be the whole animal or just some evidence that an animal was there.

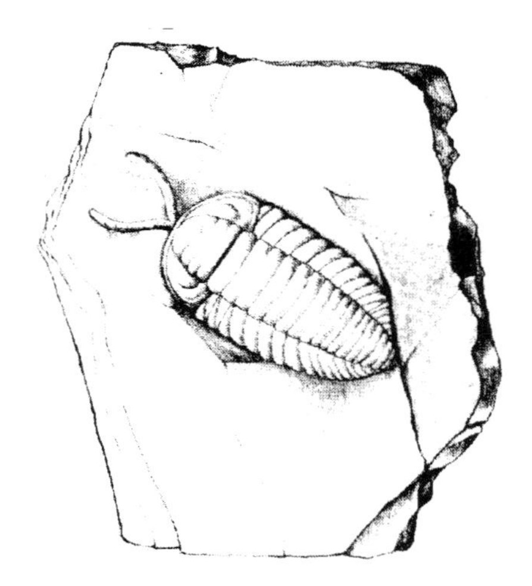

Checking Up

1. What is a fossil?
2. Give three examples of types of fossils.
3. What kind of rock is a fossil usually found in?
4. Tell how a fossil might be formed.
5. What does a petrologist do?

Teaching Suggestions The following is a list of materials needed for this learning cycle: paper cups (small), plaster of paris, stirring sticks, and a variety of small objects (coin, screw, whistle, washer, pencil).

The plaster mixture must not be too wet because, in addition to taking a very long time to dry, the imprints will not form. Tell the children to keep their imprinting objects a secret from their classmates. After the imprints dry, the children will remove the mold from the paper cup and let their classmates try to identify the object from the imprint.

You will want to check around your community to see if fossil beds are nearby. A fossil hunt can be an exciting field trip. Several sources may provide directions to fossil beds: the local/state geological survey, geologists and petrologists at a university, and local earth science teachers.

Most fossils are found in sedimentary rock. Recall that there are three main rock groups: igneous, sedimentary, and metamorphic. Rocks are classified by the way they are formed.

Igneous rocks are formed from melted rock that cools and hardens. During the cooling process, crystals (grains) of different minerals "grow." The size of the mineral grains depends on how fast the molten rock cools. Rocks with larger crystals cool more slowly than those with smaller crystals.

Sedimentary rocks are formed (1) from previously existing rocks that have been broken up into small fragments, (2) by the accumulation of the remains of living things, or (3) by chemical precipitates (dissolved material that comes out of solution). These rocks appear as fragments of materials cemented together. The fragments can be as large as pebbles or as fine as sand.

Metamorphic rocks are formed from previously existing rocks that have been subjected to heat, pressure, or hot fluids, but never melted. Pressure on these rocks produces a flattening of their minerals. Sometimes these rocks are heated by nearby molten rock. This heating can produce larger minerals in the rock formation.

Learning Cycle for Fifth Grade: Reading a Map

You can give directions to an object or a place in many ways. Using the numbers on a clock face is one way. You can use a compass and tell the direction to an object or place in degrees. Rectangular coordinates on a grid help to locate position. Landmarks help locate a particular place. In this investigation, you are going to put these ways together with other ways to determine location.

Gathering Data

Look at the map of Yellowstone National Park. Study the map closely. Where are the rectangular coordinates on the map? Where is the compass rose? Which direction is north on the map? What are some of the landmarks on the map?

Find these places on the map: Yellowstone Lake, Old Faithful Geyser, Fishing Bridge, Mammoth Hot Springs, Pyramid Peak, and West Entrance.

Tell another person how to locate each of these places. Describe the location of each place using the rectangular coordinates.

The numbers above the compass rose show the *scale* of the map. On this map, one centimeter represents six kilometers. How far is it from West Entrance to Old Faithful Geyser? Use a ruler to find the answer.

Make a map of the classroom. Use string or determine the length of a standard footstep to make a scale to find distances between objects in the room. Direct another person to an object in the room using only the map you have made. Practice finding several different objects in the room using the map. Do the same activity with a map of the school or playground.

Getting the Idea

What information were you able to find by reading a map? How is a map used to locate places on the earth? [The idea: *A map is a kind of drawing that shows the location of objects and places on earth.*]

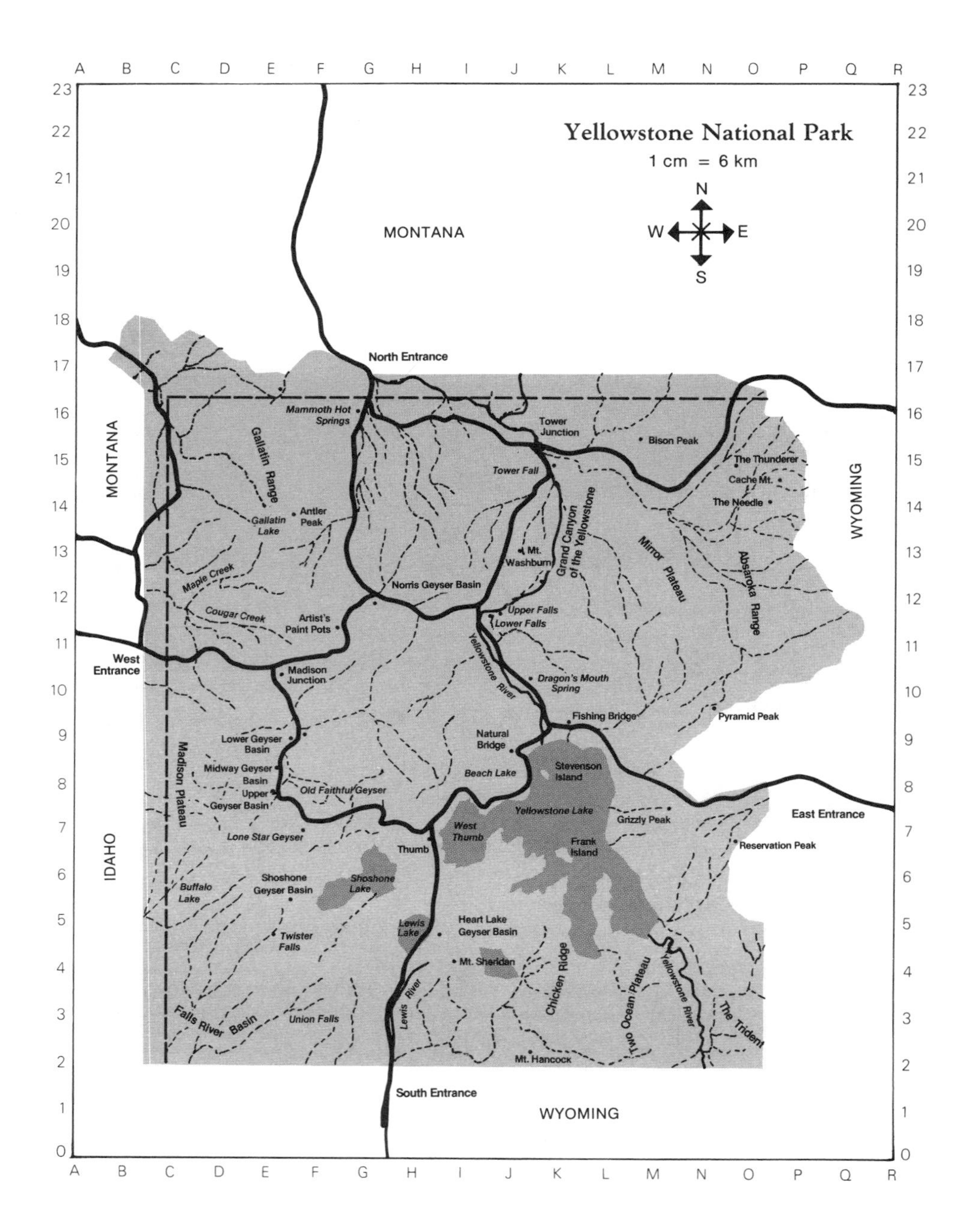
Yellowstone National Park
1 cm = 6 km
N
W
E
S
MONTANA
MONTANA
IDAHO
WYOMING
WYOMING
North Entrance
Mammoth Hot Springs
Gallatin Range
Gallatin Lake
Antler Peak
Maple Creek
Cougar Creek
Artist's Paint Pots
West Entrance
Madison Junction
Norris Geyser Basin
Tower Junction
Tower Fall
Bison Peak
The Thunderer
Cache Mt.
The Needle
Mt. Washburn
Grand Canyon of the Yellowstone
Mirror Plateau
Absaroka Range
Upper Falls
Lower Falls
Yellowstone River
Dragon's Mouth Spring
Fishing Bridge
Pyramid Peak
Madison Plateau
Lower Geyser Basin
Midway Geyser Basin
Upper Geyser Basin
Old Faithful Geyser
Natural Bridge
Beach Lake
Stevenson Island
Yellowstone Lake
Grizzly Peak
East Entrance
Reservation Peak
Lone Star Geyser
West Thumb
Thumb
Frank Island
Shoshone Geyser Basin
Shoshone Lake
Buffalo Lake
Twister Falls
Lewis Lake
Heart Lake Geyser Basin
Mt. Sheridan
Chicken Ridge
Two Ocean Plateau
Yellowstone River
The Trident
Lewis River
Falls River Basin
Union Falls
Mt. Hancock
South Entrance

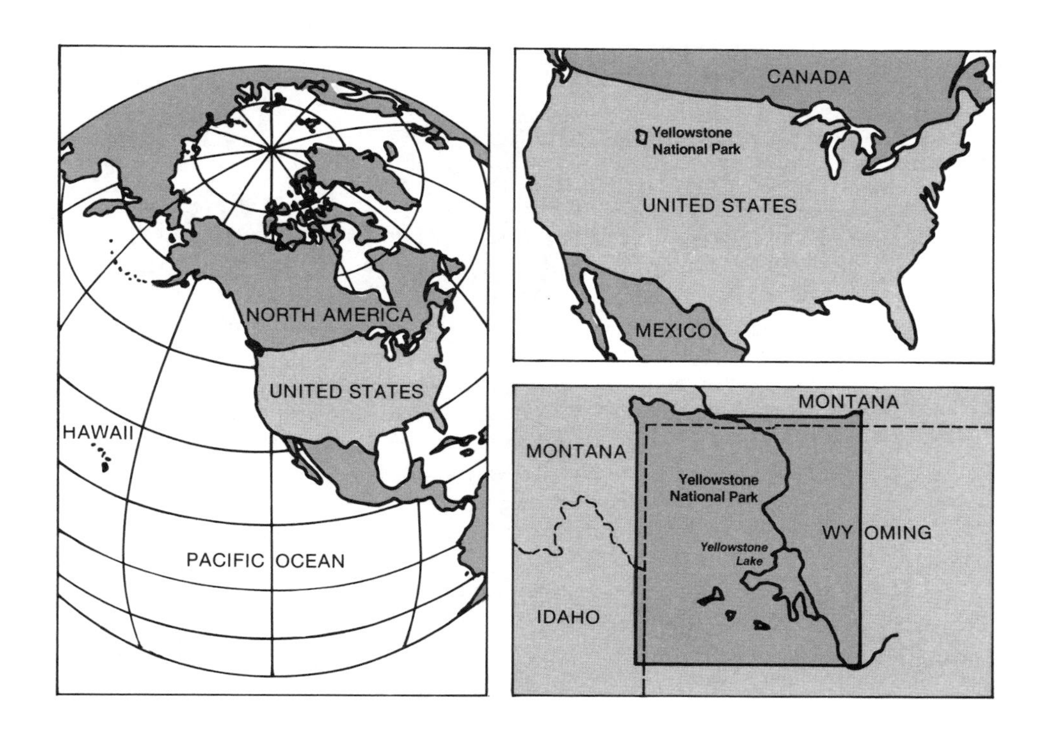
HAWAII
NORTH AMERICA
UNITED STATES
PACIFIC OCEAN
CANADA
Yellowstone National Park
UNITED STATES
MEXICO
MONTANA
MONTANA
Yellowstone National Park
WY OMING
Yellowstone Lake
IDAHO

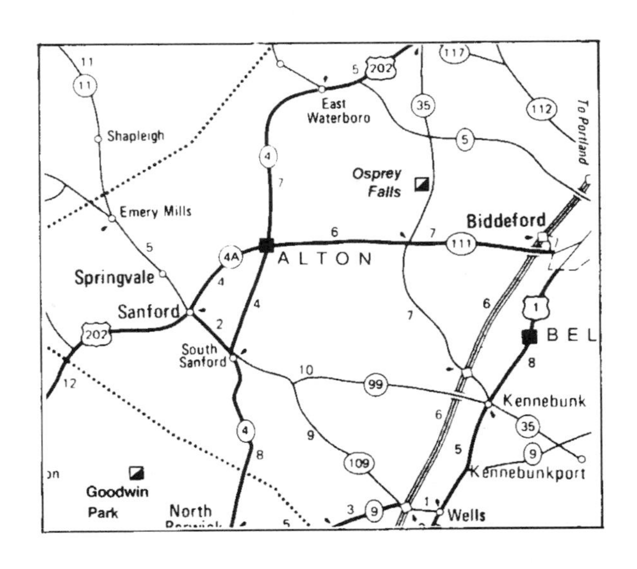
Shapleigh
East Waterboro
Osprey Falls
Emery Mills
Biddeford
ALTON
Springvale
Sanford
South Sanford
BEL
Kennebunk
Kennebunkport
Goodwin Park
North
Wells
To Portland

Applying the Idea

The scale of a map is very important. The larger the scale, the more area being shown on the map. Look at the three maps shown here. The map on the left has the largest scale. It shows all of North America and parts of Asia, Europe, and South America.

The map on the top right is a smaller-scale map. It is still a very large-scale map, however. This map shows all of the United States. The third map has the smallest scale. It shows the location of Yellowstone National Park in the western United States.

Compare each of these maps with the map of Yellowstone National Park. This map has the smallest scale. The smaller the scale of the map, the less area being shown on the map. A small-scale map shows much more detail, however.

Maps use symbols to show different kinds of objects. A *symbol* is a kind of simple picture. For example, the symbol for an airport is a picture of an airplane. Different kinds of maps use different symbols.

A *road map* is designed for people who wish to travel by automobile. It has symbols for the different kinds of roads, highways, freeways, and streets in an area. There are also symbols for points of interest, such as historical spots, and symbols for state parks, campsites, and other objects or places.

The symbols used on a map are explained in one corner of the map. All the symbols and what each one means are in a box called a *legend*. The legend for a road map is shown here.

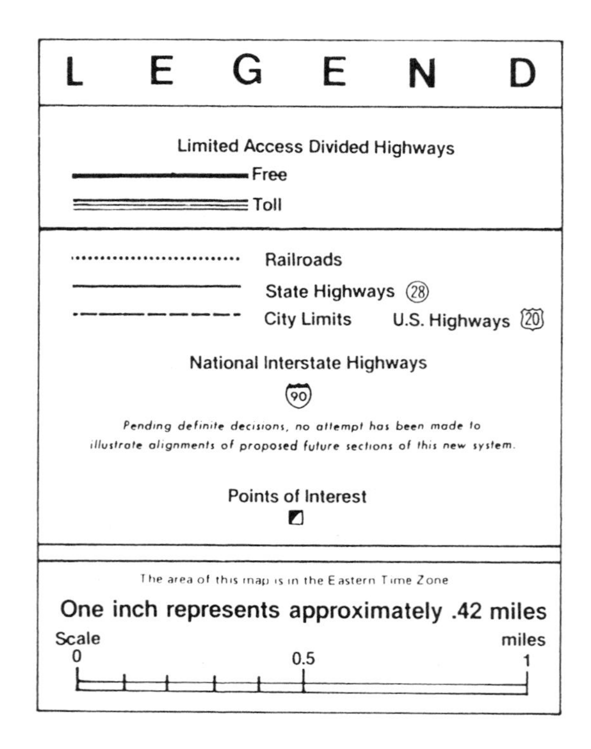

Maps also show the natural features of an area. Lakes, rivers, mountains, and other natural features are shown. Each natural feature is labeled. Sometimes the heights of the taller mountains are given.

Symbols are used to show the size of cities and towns. A small town may be shown as a colored dot. A city may be shown as a colored circle. On large-scale maps of the United States, the state boundaries are also shown.

Get a road map of your state. Examine the map closely. Find the legend. What symbols are used on this map? What is the scale of the map? What kinds of coordinates are used?

Find the location of your city or town on the map. Then choose a city or place in another part of the state. Choose a city or place you would like to visit. How many people live in this place? Check the legend to find out.

How far away is this place? Check the scale on the map. Then measure the distance from your home to this place. Use a ruler.

In which direction from your home is the place you would like to visit? Does more than one road go from your home to this place? If so, which road would you use to travel to this place? Why would you use this road?

What kinds of natural features would you pass on your way to this place? Would you cross any rivers or mountains? What points of interest are there along the way?

Write a report on how you would travel to this place. Be sure to tell how long you think the trip would take.

Examine different scale maps that show the location of the city or town where you live.

First, examine a large-scale map showing the entire country. Compare this map with a map of your state. What is the scale of each map?

Now, get a street map of your city or town. Compare the scale of this map with the scales of the other maps. Which map shows the greatest area? Which map shows the most detail? Which map would be best for finding the location of a particular street? Would this map be good for helping you plan a trip to a city in another state?

Checking Up

1. How does a map help you to find a position?
2. How can the scale on a map be used to tell distance?
3. What is the legend on a map used for?
4. What is a symbol?
5. How are natural features shown on maps?

Teaching Suggestions The following is a list of materials needed for this learning cycle: ruler (metric), state map, street map of your city, string (around one meter), and a map of the nation.

After the children become acquainted with the map of Yellowstone National Park, they will proceed with investigations relating to map reading and to finding the positions of landmarks and cities in their own state and elsewhere in the country.

Perhaps some of the children have been to Yellowstone National Park. If so, ask these students to tell about their experiences. Discuss the pictures and the references to various sites and landforms in the park.

In teaching the idea of scale, do not introduce the notion that one inch, for example, equals so many miles or kilometers. Such an understanding of scale is probably beyond the concept level of the fifth-grade child. Instead of actually defining a scale, have the children use a string. They can determine the length of string that covers the distance from one city to another. Then they can put the string over the scale. They will be able to see how many sections of the scale the string covers.

Learning Cycle for Sixth Grade: Weather Fronts

You can describe the atmosphere as you can any other object. However, there are special properties of the atmosphere that weather scientists consider important: barometric pressure, humidity, temperature, wind velocity and direction, cloud cover, and precipitation. These are elements of weather. The atmosphere, like other parts of the universe, is continuously changing. Sometimes the changes take place slowly. Sometimes they are very fast. Weather scientists look for patterns in weather changes. They also seek to explain the patterns with models. In this lesson you will learn about one of the models used by weather scientists, or *meteorologists*.

Gathering Data

Place three birthday candles on a metal tray. Place them ten centimeters apart. The base of each candle can be supported by a small piece of clay. Next, set the tray on the floor in front of a refrigerator door. Light the candles. Then open the refrigerator door. Observe the flames of the candles. What evidence can you give that cool air moves downward? Record your observations. A body of cool air such as the one that moved out of the refrigerator is called an *air mass*. In this case, it is called a *cool* air mass.

Your teacher will set up a space heater or hot plate on the floor of your classroom. The students in the class will sit in three concentric circles (or arcs) around the heat source at distances of one, two, and three meters. Each student will hold a thermometer. Before the heat source is turned on, take a reading of the air temperature with your thermometer. Record the temperature and time every thirty seconds.

Also note when you first begin to feel the heat reach your body. The body of warm air that moves out from the heat source is a *warm air mass*.

Make a class chart on the board of the time and temperature readings at the three different distances. Make a graph of the class data. Discuss and explain your data.

Getting the Idea

How did you first know when the cool air mass moved out of the refrigerator? How did you first know when the warm air mass touched your body? What did the air feel like *before* the warm air from the heater reached your body?

What were the initial temperature readings of the class? What did the air feel like after one, three, five, and ten minutes? If you stood up and walked ten meters away from the heat source, how would you know when the warm air reaches you? [An air mass on earth is a huge body of air. An air mass can be hundreds of miles wide and many miles high. An air mass has a temperature and humidity that is the same throughout. An air mass with a temperature lower than the surrounding air mass is called a *cold air mass*. If the air is warmer than the existing air, it is called a *warm air mass*.] [The idea: *When an air mass of a certain temperature moves into another air mass with a different temperature, the boundary between the two is called a front*.]

Applying the Idea

There are two common kinds of fronts. If warm air is pushing colder air ahead of it, the front is called a *warm front*. If cold air is pushing warm air ahead of it, the front is called a *cold front*.

When a warm front advances, the warm air (or light air) moves up over the retreating cold air.

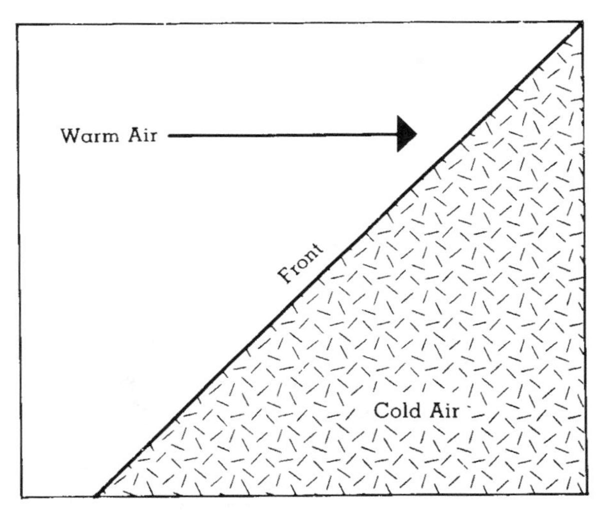

When the warm air is pushed upward by the cool air, it becomes cool. The air reaches its dew point and the moisture condenses to form large masses of clouds along the warm front. Where the level of air is highest, cirrus clouds form. Behind the cirrus clouds, stratus clouds form, with nimbus stratus, or rain clouds, next. The nimbus clouds may be very tall, building from the low to high levels of the atmosphere.

When a cold front advances, the cold air pushes under the retreating warm air. This lifts up the warm air. A cold front usually moves faster than a warm front. The heavy, cold air can move the light, warm air easily. As the warm air is lifted up quickly, the water vapor condenses to form different kinds of clouds, especially cumulonimbus (thundershower) clouds. The rains produced by a cold front are rather violent and last only a short time. These are called *thunderstorms.* Sometimes the masses of cold and warm air stop moving. The boundary between the two air masses stays in the same place. This is called a *stationary front.*

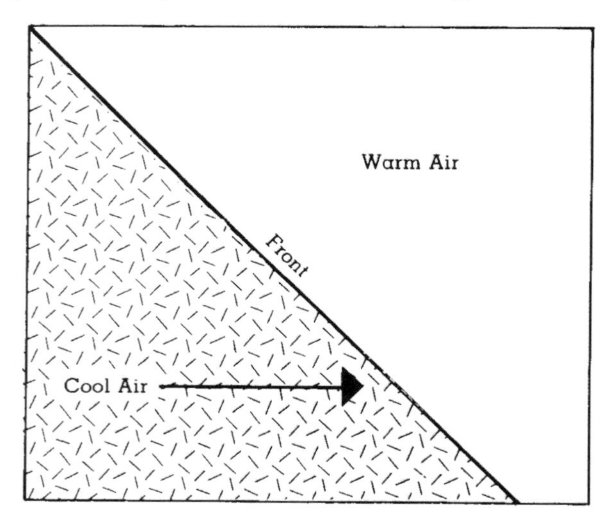

Clouds are grouped by their shape and by the height at which they are formed.

Name	
Cirrus	(Latin: curls), "horse tails," thin halos
Stratus	(Latin: spread out), layered
Cumulus	(Latin: piled up), cottonlike
Prefix	**Height or Altitude**
Cirro-	6–12 kilometers (high)
Alto-	2.4–6 kilometers (middle)
Strato-	below 2.4 kilometers (low)
Cumulo-	large vertical development

Another prefix that can be used is *nimbo,* meaning rain.

Names of clouds can be formed by combining a prefix and a name. For example, a stratocumulus cloud is a fluffy, cottonlike cloud at low altitude. Observe cloud formations during the next few weeks. Keep a record of your observations. Discuss the names you would give to different clouds you observe.

Two things that are often observed along the boundary of a front are thunder and lightning. Lightning is an electrical discharge. It can happen between clouds or within a single cloud. When lightning occurs, it heats the air, making it expand rapidly. The rapid expansion produces thunder. Observe clouds to see lightning. Is all lightning alike? Describe the different appearances of lightning. What types of clouds often produce lightning?

Checking Up

1. Describe a cold front.
2. Describe a warm front.
3. What causes clouds to form as a warm front advances?
4. What is lightning?

Teaching Suggestions The following is a list of materials needed for this learning cycle: birthday candles, clay, metal tray, thermometers, and a heat source (such as a space heater). If your students do not have access to a refrigerator at school, they can gather the data for the first experiment of this exploration at home. In the second experiment of the exploration, be sure children are made aware of dangers related to touching or tipping over space heaters.

The children will be doing experiments to observe the interaction of air masses on a small scale. Two such experiments are presented in this learning cycle. These experiments will provide the children with data that will allow them to experience models of weather fronts. For accurate results, it is important that there are no air currents in the room. If fans are operating or windows are open, the flame movement of the candles may be affected.

Take the opportunity to go outside and observe fronts and cloud types. This investigation may be conducted over several weeks. The fall and spring season will provide a continuous variety of weather fronts and cloud formations.

Learning Cycles for the Physical Sciences

Six concepts from those listed at the beginning of this chapter have been selected for the term introduction phases of the learning cycles that follow. These concepts are the following:

Grade 1: Color, shape, size, feel, and smell are properties of *objects* (physical science concept #2).

Grade 2: *Magnets* are special objects that stick to or attract iron (physical science concept #8).

Grade 3: The three states of matter are solid, liquid, and gas (physical science concept #1).

Grade 4: Matter that has variable properties is called a *mixture* (physical science concept #4).

Grade 5: *Energy* is something that causes objects to do things they would not do without it (physical science concept #1).

Grade 6: The interaction of the wire, the dry cell, and the lightbulb causes *electrical current* in the circuit (physical science concept #10).

Learning Cycle for First Grade: Properties

Name all the objects. How are the objects different? We can find out.

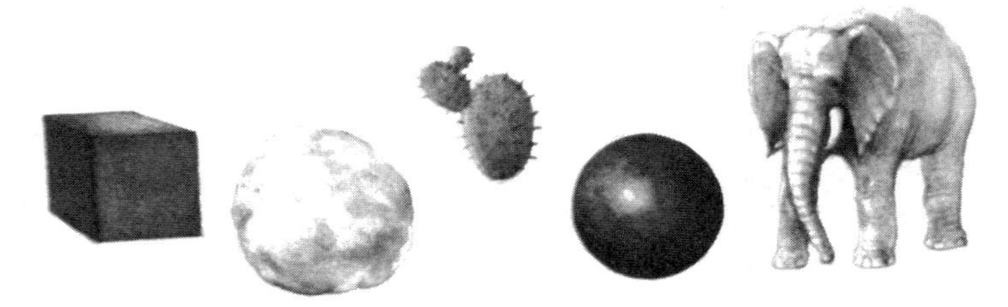

Gathering Data

Name the color of the top. [The objects should be different colors when student sheets are prepared.] Name the color of the whistle.

The objects have the same color. How are they different?

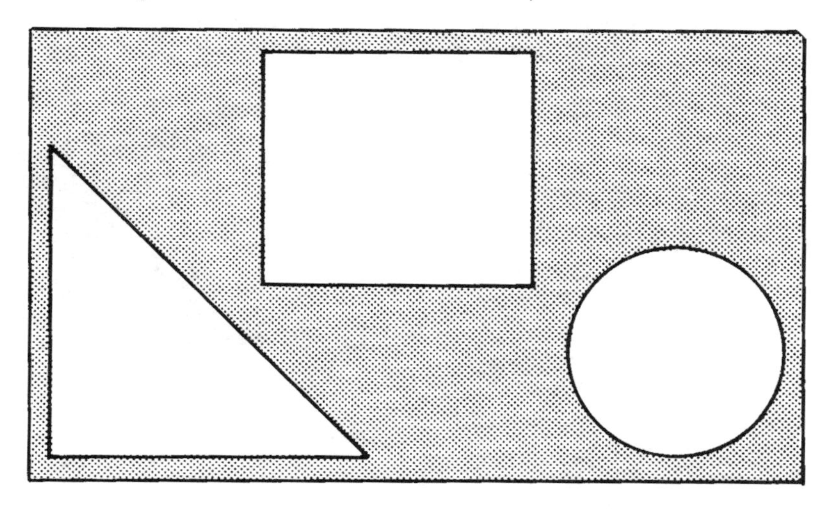

Name all the objects. Each object has a shape. What shapes do you see?

Look at the mother hen. See the baby chicks. The mother hen is big. The baby chicks are small. Look at the mother dog. See the little puppy. The mother dog is big. The puppy is small.

Roll some clay into a ball. Make a big, round ball. Make another ball of clay. Make a little ball.

Go to the window. Rub the glass. Rub it with your hand. How does the glass feel?

Get a piece of sandpaper. Rub the sandpaper. Rub it with your hand. How does the sandpaper feel?

Get a piece of wool. Rub the piece of wool. How does the wool feel?

Getting the Idea

What can you tell about objects you see? What can you tell about objects you touch or smell? [The idea: *Color, shape, size, feel, and smell are properties of objects.*]

Applying the Idea

The objects have properties. Each object has two of the same properties. Name the two same properties.

Checking Up

1. Here are things in rows.
2. What objects do you see?
3. What objects do you see in each row?
4. Which object in each row does not belong?

Teaching Suggestions The following is a list of materials needed for this learning cycle: numerous objects of various colors; two or three colored tops; two or three colored whistles; construction paper of various colors; classroom objects of various shapes; cardboard shapes (triangles, squares, circles, rectangles); cotton balls; numerous small, hard objects; numerous small, soft objects; pieces of cloth; pieces of wood; and sandpaper.

In this learning cycle, the children will be conceptually developing the concept of *property.* (Property is the most basic concept of all science.) Explain that color is a property, size is a property, and feel (texture) is a property. Stress that properties *tell* about an object.

Pick up one object and ask a boy or girl to identify its properties. Do not accept as properties the name of the object or what it does. Explain that a name or what an object can do does not tell anything about the object. *Property* is the thing that tells about an object. Property tells what an object is like.

Point out that an investigator can determine color, shape, and approximate size by looking at an object. But to know about texture, an individual must feel the object. Explain that words such as *hard, soft, rough, smooth, sharp, dull,* and *bumpy* can be used to describe texture.

You might reinforce the idea that *name* and *use* are not properties by first talking about the name of an object and discussing its use. For example, you could talk about a pencil. Ask the children to give the name of the pencil and tell about its use. Listen to their responses. Then say, "Now, what are the pencil's properties?"

Learning Cycle for Second Grade: Magnetism

There are many metals in the world. Gold is a metal. Steel is a metal. Silver is a metal. Iron is a metal. Metals help us. Steel is used in cars. Gold is used as money. So is silver. Iron is used in many objects. You are going to experiment with iron.

Gathering Data

You need iron nails. You need two other iron objects. Touch one iron object to another. What happens? Try to make the iron objects stick together. What happens? What do the iron objects do to each other? Get a special iron object. It is shaped like a bar. Get it from your teacher. Use the bar-shaped object. Use it with other iron objects. Tell what happens. Use objects not made of iron. Use the bar-shaped iron object. What happens?

Here is a different object. Get one like it. Use the new object with iron objects. Tell what happens. Use the bar-shaped object from the last experiment. Use it with your new object. Tell what happens. How is the new object like the bar-shaped object? How are the objects different?

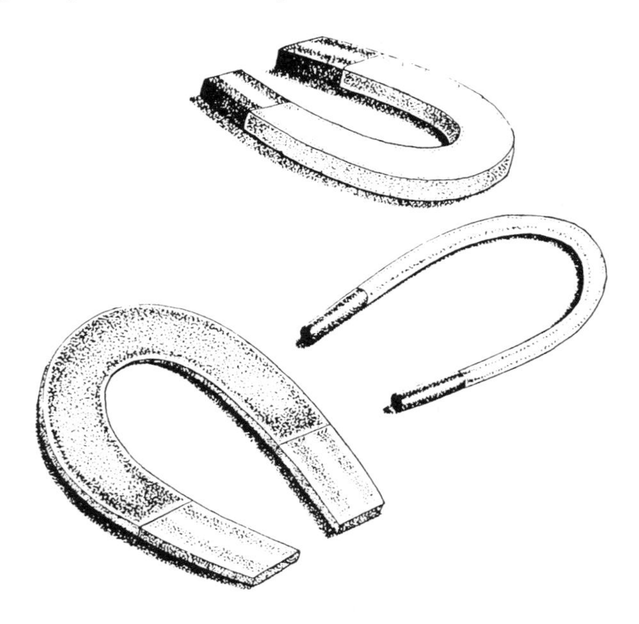

Getting the Idea

What did you observe when you touched the metal bar to the iron? What did you observe when you touched the u-shaped object to the iron?

What did you observe when you touched the metal bar to objects that were not made of iron? What did you observe when you touched the u-shaped object to objects that were not made of iron? [The idea: *Magnets are special objects that stick to or attract iron.*]

Applying the Idea

Get a magnet. Be sure it is a good one. Test it. Attract paper clips with it. Use many other objects. Which ones does the magnet attract? What materials are the objects made of? Which ones does it not attract? What materials are the objects made of? What materials do magnets attract? What materials do magnets not attract?

Checking Up

1. What does a magnet do?
2. What were the shapes of the magnets you experimented with?
3. Magnets attract each other. What does that mean?
4. Magnets attract iron objects. What does that mean?
5. What materials do magnets not attract?

Teaching Suggestions The following is a list of materials needed for this learning cycle: various types of magnets and objects for the class to test magnetic attraction, for example, chalk, marble, nail, coat hanger, eraser, nuts, and bolts.

Your students will be investigating the properties of a magnet. They will be developing the concept of *magnetic attraction*. Try to have various kinds of magnets available for the class.

During the Getting-the-Idea phase, use the term *magnetic attraction* and continue to use it as appropriate. Proceed by telling the students that they will be finding out what other objects show magnetic attraction. Among objects they might test are a pencil, eraser, marble, paper clip, metric ruler, crayon, piece of chalk, ballpoint pen, nail, nail file, and a coat hanger.

Explain to the class that they need to keep a record of what does and what does not show magnetic attraction. Encourage the children to test many objects and to make a record of each test.

You might put a list of the objects the children test on the chalkboard. The class can discuss the list after the students have completed their testing. In the discussion, ask the students to explain how the items that showed magnetic attraction seem to be alike. What property do they have in common?

If the children are uncertain, spend some time naming some of the properties of the objects. Be sure the children name the materials that have been tested. If the term *metal* does not come up, introduce the term yourself. Lead the children to note that all the objects that showed magnetic attraction are made of some kind of metal.

Learning Cycle for Third Grade: States of Matter

Suppose you throw a glass of water on a window. The water will splash and run down the window. Try it and see. Suppose water is frozen into ice. Suppose you threw the chunk of ice against the window. What would happen? Don't try it!

A liquid, like water, is something that can be poured. When you pour a liquid into a cup, it takes the shape of the cup. You cannot pour a solid. A solid has its own shape. To change the shape, you must break it apart. But what is a gas? You can find out by doing an experiment.

Gathering Data

Wet a piece of cloth in water. Squeeze the cloth to remove most of the water. Put the wet cloth in front of a fan. Observe carefully as the cloth dries. Where does the water go? Write down what you believe.

Water is boiling in a pan. Steam rises from the pan. Observe the steam. Your teacher holds a cool spoon above the pan. What collects on the spoon? How is this like the fan and the wet cloth? Write down an answer to this question.

Put water in a jar. Fill the jar almost to the top. Then put ice in the water. Put a lid on the jar. Next, wipe the outside of the jar. Wipe it dry with a paper towel. Observe the outside of the jar for several minutes. What collects on the sides of the jar? Explain what happens. Make a record of the experiment. How is this like the steam and the spoon experiment? How is it like the fan and the wet cloth experiment? Write down what you think.

Place an ice cube on a plate. Set it on a window sill. Put it in the sun if you can. What happens to the ice? Write down what you found.

Put some water in a pan. Put the pan in a freezer. Leave it for a day. What happened to it? Make a record of what you found.

Getting the Idea

What did the liquid water change to when it was frozen? What did the solid ice change to when it melted? What did the liquid water change to when it boiled?

Look at your responses to the last three questions. What three different forms, or *states,* can matter (such as water) be found in? [The idea: *The three states of matter are solid, liquid, and gas.*]

Applying the Idea

As you have observed, a liquid can be frozen. Water freezes into ice, a solid. When the water freezes, there is a change in it. There is a change in how it looks and feels, or in the way it is. This is called a *change in state.*

To *evaporate* is to change to a vapor. This happens when water turns to a gas. A solid can change to liquid. This happens when ice melts. Water is matter. It can be in three states. Water can be a solid, a liquid, or a gas.

Water vapor *condenses* on a cool surface. To condense is to change from the gas state to the liquid state. There is water vapor in the air all the time. Usually you cannot see it. But you can see water vapor when it condenses because it turns into liquid water.

Maybe you have gone for a walk early in the morning. When you walked through grass, your feet or shoes got wet. The grass had water on it. It was wet. There had been no rain. Still, the grass was wet.

When night comes, the earth gets cool. Plants on the earth get cool, too. Water vapor in the air turns into drops of dew on the plants. The water vapor condenses and turns into liquid water. The water goes from the gas state to the liquid state.

There is dew on the grass early in the morning. Then the sun comes up. It shines down on the plants and the dew. The sun warms them. The dew changes from the liquid state to the gas state. It rises into the air as water vapor. The dew *evaporates*, or changes into a vapor. The process is known as *evaporation*. Matter goes from the liquid state to the gas state.

Fill an aquarium tank with water. Then lower a glass into the water. Put the glass in the water with the open end down. Observe what happens. What happens to the air in the glass? What two states of matter are here?

Now, tip the glass. Explain what happens. What did one state of matter do to the other?

Now, let us investigate air. First, catch some air in a plastic bag. Fill the bag with air. Close the bag by twisting the open end. Keep the air from escaping. Observe the air carefully.

How does air differ from water, a liquid? What state of matter is air? How does air differ from ice, a solid? Make a record of the differences. List some properties of air.

Checking Up

1. Look for the three states of matter in your classroom. Name them.
2. How is the gas state of matter different from the liquid state?
3. Drops of water collect on a glass filled with ice. What change of state of matter was observed?
4. How can you change the state of matter of water?
5. How is the liquid state of matter different from the solid state?

Teaching Suggestions The following is a list of materials needed for this learning cycle: aquarium tank or any large container with water in it, wiping cloths, drinking glass, fan, ice, jar with lid, plastic sandwich bag, spoon, hot plate, and container to boil water.

The first experiment in the Gathering Data section enables the children to observe evaporation. You might prefer to do this experiment as a demonstration. The fan will hasten the evaporation of the water in the cloth. Have your students examine the cloth when it has dried out. Explain that the water in the cloth evaporated, or changed into a gas.

An experiment with boiling water provides another example of evaporation. Use a hot plate to boil some water in a beaker or teakettle. Explain to the class that, when it boils, water changes into water vapor, a gas. Call attention to the steam. Point out that the visible steam is not water vapor. The steam consists of water droplets that condensed from the water vapor. The cool surface of the spoon caused the water vapor to condense.

The next experiment provides another example of condensation. It is a good idea to do this experiment during the same period in which you boiled the water and observed evaporation. The two experiments go together.

Have the children work in groups of three or four to do the experiment with the ice water. Let each group fill a jar with ice and water. Instruct the children to seal the lid on the jar tightly. Tell them to be sure to wipe the outside at the beginning of the experiment. Have them form a circle around the jar to observe the results of the experiment.

In time, the water in the air will condense on the cool surface of the jar. Have the children discuss the experiment. In all likelihood, most of the students will conclude that the water that collects on the jar came out of the air. If some of the children have the notion that water seeped through the jar, repeat the experiment using warm water. Let the children discuss the experiment.

In the experiment with air, the children will observe that air takes up space. Air is matter. They also will observe that air takes up space when they put a drinking glass into the water in an aquarium tank.

The students' experiments and observations have made it clear to them that matter exists in states. Introduce the term *states of matter*. In your discussion, call upon students to explain the difference between a liquid and a solid and between a liquid and a gas.

Learning Cycle for Fourth Grade: Mixtures

A substance is matter that has definite properties. Substances can be elements, such as copper, oxygen, gold, or mercury. Compounds such as salt, water, and sugar are also substances. Each piece of matter just named has definite properties. It is a substance. You will now investigate another kind of matter.

Gathering Data

In this experiment you will mix one substance with another. You will stir them together. In each part of the experiment you will control the amount of one substance. You will vary the amount of the other substance.

Use five glasses of tap water filled to the same mark. Fill each glass almost full. To each glass add the following amounts of table salt and stir:

Glass 1: one small crystal
Glass 2: one-quarter of a teaspoon
Glass 3: one-half of a teaspoon
Glass 4: one teaspoon
Glass 5: one tablespoon

Stir the salt and water until all of the salt disappears. Then look at the salt-water systems. Observe each salt-water system. If the salt-water systems look different, record that difference.

Next, taste each salt-water system. Compare the taste of each system. What variation do you notice? What was the controlled variable in this experiment?

Repeat the experiment you just completed. In repeating the experiment, use food coloring instead of salt. Put one drop of food coloring in one glass. Put two drops of food coloring in another. Put three, four, or five drops in the other three glasses. Stir the water and the food coloring. Describe the differences in color from one glass to another.

Stir some solids together. Put table salt and colored sand together. Stir them well. Make several combinations of salt and sand. Vary the amount of one of the substances in each of your combinations. Observe the systems with a hand lens. Describe what is different about each system.

Getting the Idea

What property did you observe to be *different* or *variable* among each of the five glasses in the salt and water experiment? Why was this property different or variable among each of the glasses?

What property did you observe to be *different* or *variable* among each of the five glasses in the food coloring and water experiment? Why was this property variable among each of the five glasses?

What properties were *different* or *variable* about each system of salt and colored sand you made? Why were these properties variable? How would you summarize the variations you observed in the above experiments? [For each experiment, the taste, color, and/or texture varied among the different systems. Taste, color, and texture are properties that can change when a different amount of one substance is added to a system.] [The idea: *Matter that has variable properties is called a mixture.*]

Applying the Idea

You can put together different amounts of salt in water. That is, you can vary the amount of salt. The salt-water systems look very much alike. But there is a variable. Their taste is variable.

When food coloring is mixed with water, the color of the systems varies. You mixed sand and salt. Again, you observed a variation in properties.

Most of the materials you use are mixtures. Many foods are mixtures. Soft drinks are mixtures. Some of the materials you find in the kitchen are pure substances, such as sugar, salt, and baking soda. But you almost always mix these together with other materials before you sit down and eat. Cooking can change a mixture to one or more compounds. But then the new compounds make up a mixture.

Air is a mixture. It has elements mixed together in it. Those elements are oxygen, argon, and nitrogen. Carbon dioxide is also in the mixture called air. Carbon dioxide is made of two elements, carbon and oxygen. It is a *compound.*

There are also substances in the air mixture. All air has some dirt in it. Most furnaces have objects that take out most of the dirt. Those objects are called *filters.* Breathing a dirt-and-air mixture can be unhealthy.

The properties of a clean-air mixture make air healthy to breathe. Remember that the properties of a mixture depend upon the amount of each substance in it. Too much dirt in the air mixture makes it unhealthy. There are many other substances in air. Automobiles give off a compound called carbon monoxide. This compound is poisonous. Breathing too much of it can cause sickness or death. In cities where there are many automobiles, the properties of the air mixture change. The properties of the air mixture become more and more like carbon monoxide. As more salt is added to water, the mixture tastes less like water and more like salt. That is what happens to air. The air mixture becomes *polluted.*

Not all air pollution comes from automobiles. Many factories also let harmful materials escape into the air. When that happens, the properties of the air mixture change. They become less like a clean-air mixture. The properties of the air mixture become more and more like the harmful substance entering it. Air pollution must be stopped. If it is not, we won't have a clean, healthy air mixture to breathe.

Mixtures can be separated into substances. One way to separate a mixture of water and another substance is by heating. Water evaporates, or changes to a gas. The gaseous water can then be condensed, or changed back into a liquid.

Water that is evaporated and then condensed to a liquid again is called *distilled water.* Almost pure water can be obtained by distilling water. When the water is distilled, the dissolved material is left behind.

Make a simple distilling apparatus. Set up a flask, a beaker, a hot plate, and some glass tubing. Separate the salt and water from a salt-water mixture.

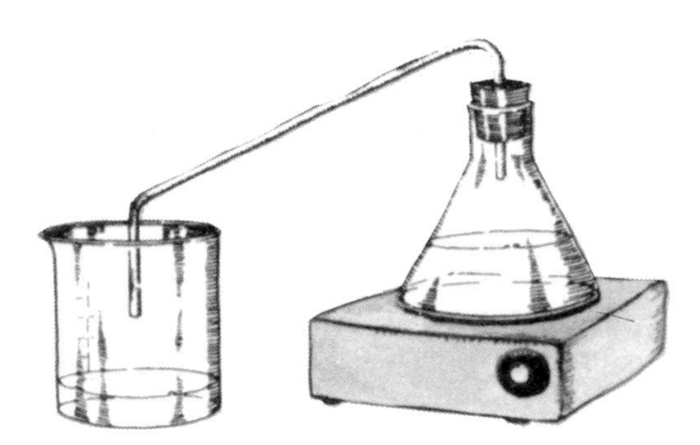

Mix a teaspoon of salt with a teaspoon of clean white sand. Use a hand lens to observe the mixture. Then place the mixture in an empty tea bag and staple it closed. Place the tea bag on the side of a glass. Put water in the glass. Put enough water in the glass to cover about half of the tea bag.

Observe the interaction. When no more of the mixture dissolves, remove the tea bag. Dry the material left in the tea bag and examine it. Explain why some material is left in the tea bag. How did the properties of the water change? What is the water now?

What property does salt have that the sand does not have? Tell how this variation in property allows you to separate salt and sand.

Try separating another mixture. This mixture is chalk and salt. First, mix together a teaspoon each of chalk and salt. Put some of the mixture in an empty tea bag and staple it closed. Put the tea bag in a glass and add water. Observe carefully and describe the results.

Add one cubic centimeter of white vinegar to a glass that contains a mixture of water and bromothymol blue (BTB). Observe any change you see in the mixture. Add another cubic centimeter of vinegar and observe any change in the mixture. Keep adding BTB one cubic centimeter at a time. Compare the properties of the BTB-water mixture with the properties of the BTB-water-vinegar mixture. Record your comparisons.

Checking Up

1. What is a mixture?
2. Tell why the vinegar, BTB, and water you put together form a mixture.
3. When a person washes dishes or clothes, a mixture is used. How is the mixture made?
4. Lemonade is a mixture. How does one lemonade mixture vary from another?
5. How could you separate a mixture of sand and iron fillings?
6. How can you change the properties of a mixture?
7. How is air pollution caused by a mixture?

Teaching Suggestions The following is a list of the materials needed for this learning cycle: beaker, bromothymol blue (BTB), chalk dust, flask with rubber stopper for glass tubing, food coloring, funnel, glasses, glass tubing approximately thirty-five centimeters long, hand lens, hot plate, ice or cold water, measuring spoons, medicine dropper, paper plates, salt, sand (colored and white), stoppers (one-hole), sugar, tea bags, and vinegar.

The first three experiments in this learning cycle established a key idea: You can vary the amount of the substances that are put together to form mixtures. In a mixture, the substances that are mixed together retain their own properties. There is no chemical change in any of the substances when they are combined. Be sure the children understand that a mixture consists of two or more substances.

Distillation is the process commonly used to separate mixtures in a laboratory. You can refer to the illustration in the learning cycle as a guideline for setting up the apparatus. First, bend the ends of a length of glass tubing. You can do this by heating the glass over a flame. Insert one end through the hole in a one-hole rubber stopper. Add a portion of your salt solution to a flask. Put the rubber stopper with the glass tubing into the mouth of the flask. Place the other end of the glass tubing in a beaker.

Put the flask on a hot plate. As the solution boils, water vapor will escape through the glass tubing. The beaker should be set in a pan of ice or cold water. The water will condense as it drains into the beaker. Liquid water will be collected in the beaker. Salt crystals will remain in the flask as a *precipitate*. Since you will be boiling water, you might do this experiment as a demonstration.

Expect the following to occur in the experiments suggested under Applying the Idea: the salt in the tea bag dissolves in the water that soaks through the tea bag. The dissolved salt then diffuses through the tea bag into the glass containing water. The sand remains in the tea bag. Salt has the property of *solubility*. Sand does not dissolve in water. The fact that salt dissolves makes it possible to separate the salt from a salt-sand mixture.

Salt can also be separated from chalk by filtering the solution. The liquid that passes through the filter is called the *filtrate*. The solid that remains on the filter paper is called the *residue*. The solid is chalk; chalk is the residue. Salt passes through the filter along with the water. The salt could be recovered by distilling the water.

In the experiment with vinegar, the baking soda will fizz. The vinegar contains acetic acid. The acetic acid reacts with the baking soda, causing carbon dioxide to be released.

Materials can also be tested with bromothymol blue (BTB). The BTB causes a change in color. As the children will observe, a solution of BTB and white vinegar is yellow. A solution of BTB and ammonia water is blue.

Learning Cycle for Fifth Grade: Energy

A car speeds along the freeway. The traffic lanes are jammed. The rush-hour traffic slows to a crawl now and then. But in the open spaces, the cars are moving at the legal speed. The cars have *motion.*

A speeding train has motion, too. A waterfall and a whistling teakettle have motion. A roller skater and a bicycle rider have motion.

The car, the train, the waterfall, the teakettle, the roller skater, and the bicycle rider all have something in common. True, they all have motion. But what else do they have in common? What causes them to have motion?

Gathering Data

Build a ramp like the one shown in the picture. Be sure one end is at least ten centimeters higher than the other end.

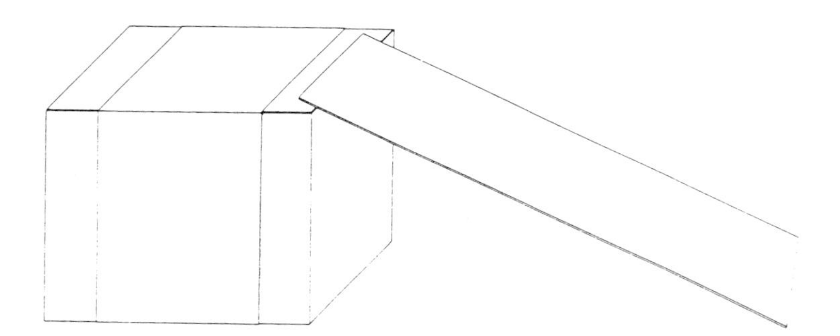

Place a paper cup at the bottom of the ramp. The open end of the cup must be toward the ramp. You now need a large sphere. A golf ball will work nicely.

Place the golf ball at the top of the ramp. Let it roll down the ramp. Be sure it rolls into the cup. Record what you see.

You need a large metal washer. Tie a string around it and make a pendulum. Set the paper cup on the table. Hold the string in one hand so the washer hangs straight down. Use your other hand to pull the washer back as far as you can. Let the washer go and strike the cup. Watch the cup. Record what happens.

Getting the Idea

What did you observe happen to your cup in the ramp experiment? Why did this happen to your cup?

What did you observe happen to your cup in the pendulum experiment? Why did this happen to your cup?

Compare the two experiments. In each experiment, what caused the objects (cups) to behave the way they did? What would have happened to the objects if you left them alone? [The idea: *Energy is something that causes objects to do things they would not do without it.*]

Applying the Idea

When moving objects strike objects at rest, something always happens. If you ride a bicycle into a large cardboard box, you know something will happen. If you throw a ball into a catching net, you can see something happen.

When water falls and strikes the earth something happens. The earth is eventually worn away. If a glass object is held up high and then dropped, something happens. The glass object is probably broken. Imagine that a rubber ball is held above your head and dropped to the floor. Something happens. The ball bounces back up. All of these events happen because of energy. *Energy is something that causes objects to do things they would not have done without it.*

When a ball is thrown at a batter it can go right past. If the ball does not pass the batter, energy must be used. The batter must use energy to swing the bat. Suppose the moving bat strikes the ball. The ball goes back toward the pitcher. The ball would not have done that if the bat had not hit it. In order for the bat to hit the ball, the batter must use energy. Energy causes something to happen that does not happen without it.

Suppose you have been told to wash a car. The soapy water will just stay in the pail and the dirt will stay on the car. But you can use energy to scrub the dirty car with the soapy water. The car then becomes clean. In order for that to happen you had to use energy.

You will use the ramp that you built earlier to do another experiment. You will need three different size spheres. A baseball, a golf ball, and a marble will do.

Place the paper cup at the bottom of the ramp as you did before. Then let one of the spheres roll down the ramp into the cup. Measure the distance the cup moved. Repeat this for the other two spheres. Which sphere pushed the cup farthest? Which sphere had the most energy?

Again, use the ramp that you built earlier. This time use only the golf ball. First, set the box on the table as you did for the other experiments. Roll the golf ball down the ramp into the paper cup. Measure how far the cup moved. Now, raise the higher end of the ramp. You can do this by placing books under the box. Again, roll the golf ball down the ramp. Be sure to start the ball at the same place as before. Which time did the cup move farther? Which time did the ball have more energy?

Use the pendulum with one washer that you used earlier. Set your cup on the table and swing the washer toward it. Measure how far the cup moved.

Now put three washers on the string and repeat the experiment. How far did the cup move? Which pendulum provided the most energy?

Checking Up

1. What is energy?
2. You guide a power lawn mower to mow the lawn. Where does the lawn mower get its energy?

3. A golf ball travels down the fairway. Where does it get its energy?
4. What makes it possible for a bowling ball to knock over bowling pins?
5. Look at the pictures below. Where do the Ping Pong ball, the Frisbee, and the wagon get energy?

Teaching Suggestions The following is a list of the materials needed for this learning cycle: baseball, boards (ramps), paper or plastic cups, golf ball, spheres (metal, marble-sized), string, and weights (or bolt nuts).

The children will begin by doing an experiment with an inclined plane. The sphere pushes the paper cup. The children may also observe that, the higher the ramp, the greater the distance the paper cup is pushed.

Refer to the idea, *the experiments you have done produced the results they did because of energy*. This is probably the most significant concept that the children will encounter in their study of fifth-grade science. *Energy* is what makes things go. For your students, this is a much better definition of energy than the conventional definition, which is that energy is the ability to do work. To understand the formal definition, the student must know the meaning of the word "ability" and must be able to explain the word "work." Without such prerequisites, the child can develop a concept of energy merely by understanding that *energy is what makes things go*. Teach this definition to your class.

Learning Cycle for Sixth Grade: Electricity

Electricity is energy. A special system is needed to use electrical energy. No matter what is using electrical energy, that special system is there. That special system follows certain rules. In this lesson you will find out about some of those rules.

Gathering Data

You will need a wire, a dry cell, and a lightbulb. The objects are shown in the picture. Put the three objects together. What evidence do you find that the objects will interact? After you find evidence of interaction, draw a picture of how you connected the wire, the dry cell, and the lightbulb.

First, use one wire to produce interaction. Then use two wires. Draw a pic-

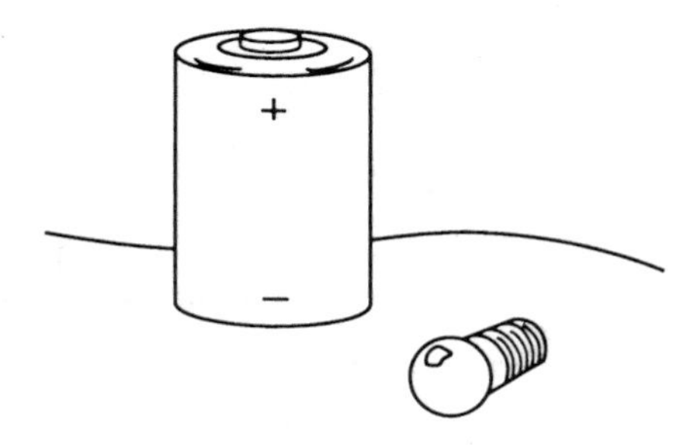

ture of how the two wires, the dry cell, and the lightbulb were connected when you saw evidence of interaction.

Compare the two pictures you have drawn. How are they alike? How are they different?

At how many *different* points did you touch the dry cell before you observed evidence of interaction? At how many places did you touch the lightbulb before you saw evidence of interaction?

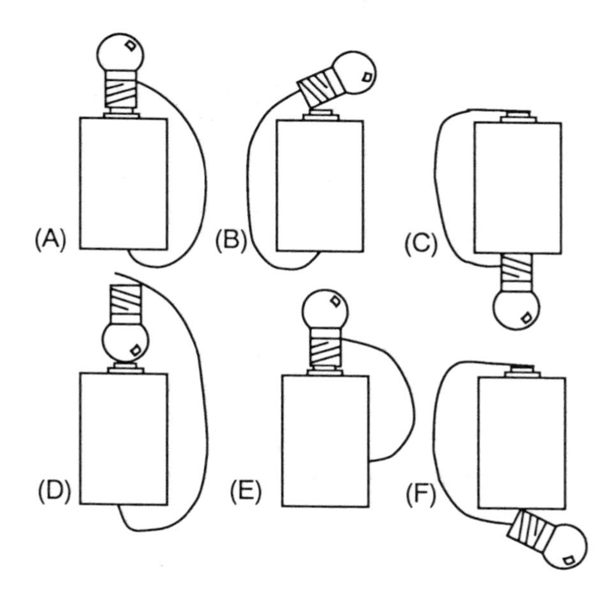

Each of the six pictures on this page shows a system made up of a wire, a dry cell, and a lightbulb. The objects are connected differently in each picture. Predict in which of the systems the bulb will light. Record your predictions in a table like the one shown below. Write "yes" or "no" in the "Prediction" column.

From now on, do not change any of your predictions. You are going to use a wire, a dry cell, and a lightbulb to test your predictions. Arrange the objects

Picture	Prediction (yes or no)	Experiment (yes or no)
(A)		
(B)		
(C)		
(D)		
(E)		
(F)		

just as they are shown in pictures A, B, C, D, E, and F. Record in your table whether or not you found evidence of interaction by writing "yes" or "no" in the "Experiment" column.

Getting the Idea

When you arranged the wire, dry cell, and bulb in a certain way, what interaction did you observe? Describe how you arranged these three objects to produce this interaction. [We made a *circle* or *circuit:* One end of the dry cell touched one end of the wire, the other end of the wire touched the metal part of the bulb, and the metal end of the bulb touched the opposite end of the dry cell.]

When you arranged your objects in the manner described above (in a circle or circuit), what was present within the system that *caused* the interaction? [The idea: *The interaction of the wire, the dry cell, and the lightbulb causes electrical current in the circuit.*]

Applying the Idea

A wire, a dry cell, and a lightbulb can be arranged in a system that does something. There is evidence of interaction. The lightbulb lights.

With the lighting of the bulb, energy is being used. The dry cell supplies the electrical energy. But in order to see evidence of interaction, you had to arrange the wire, the dry cell, and the lightbulb in a certain way. With that arrangement, you observed evidence of interaction. Such an arrangement is called an *electrical circuit.* The energy within an electrical circuit is called *electrical current.*

The word *current* means anything that is running or flowing. Maybe you have watched a large crowd when you were looking down from the window of a tall building. The entire crowd seems to stand still but the people in it move or flow through the crowd. This is a current of people. The people flow along through the crowd.

You are probably familiar with a current of water. Suppose it's a river. The river seems to stand still, but the water moves. Water moves along a certain path. That movement is often said to be a "swift current" or a "slow current." Persons using these phrases are using the true meaning of the word current. The water is flowing along.

When giving weather reports, weather forecasters often talk about currents. They discuss air currents. Differences in temperature cause air currents. These currents can affect the type of weather we have. When air currents get moving too rapidly, we call them wind.

You know that the dry cell or battery of cells in a circuit is the energy source. The lightbulb or some other object is the energy user. But in order for a lightbulb to use energy from a dry cell, the energy must get to it. The wires are the connectors in an electrical circuit. The lightbulb does not light until wires connect it and the dry cell.

When the bulb lights you have evidence that the bulb, wire, and dry cell are interacting. The evidence of the bulb lighting can be thought of as showing something additional. The light in the bulb suggests that the wires are carrying energy from the dry cell to the bulb. The wires are actually transferring the energy.

When you think of the evidence in that way, it is not hard to see why the name *current* was used. The electricity flows just as people flow through a crowd. The flow of electricity is like the flow of water in a river. So the flow of electrical current is used to discuss what happens in the wires. But remember, it is a model. No one has ever seen it happen.

To make the model more clear, some people call the electricity "juice." This model says that something that flows like juice is present in the wires. Again, remember that the flow of electrical current is a model. This model explains how the electrical energy gets from the energy source to the energy user.

Let's do another experiment. You need a dry cell, a lightbulb, two brass clips, three pieces of wire, a holder for the dry cell, and a socket for the lightbulb.

Put the objects together as shown in the drawing. As you will observe, the lightbulb gives no evidence of interaction. Next, touch wire A and wire B together. What happens?

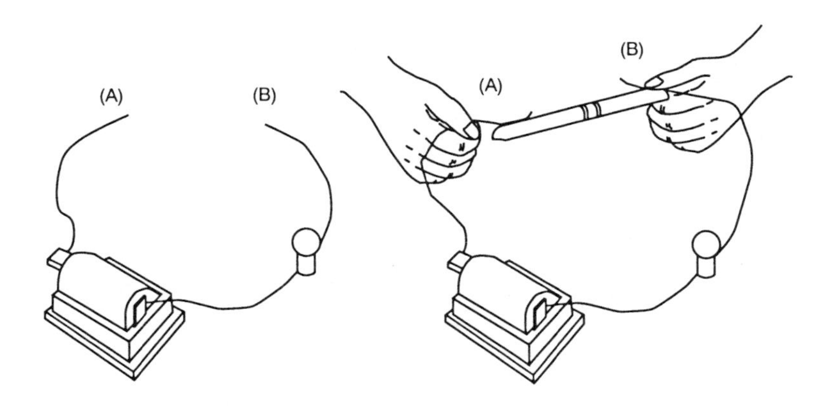

Put different kinds of objects between wire A and wire B. Touch the two wires to the objects. Which objects cause the lightbulb to light?

When the lightbulb lights, the circuit is a *complete circuit*. Make a record of the kinds of materials you used in your circuit. Indicate which materials made a complete circuit.

Refer to the drawing above. When a system of electrical objects is arranged like the system in the first photograph, the arrangement is called an *open circuit*. The circuit is a *closed circuit* when something is put between wire A and wire B and the lightbulb lights.

The system you just worked with is a *circuit tester*. A circuit tester quickly tells you when you have a complete circuit. You now can do an experiment with a circuit tester.

Push two brass paper fasteners through a piece of cardboard. Put them five centimeters apart. Connect the two brass fasteners with a piece of wire. Connect them on the back of the card.

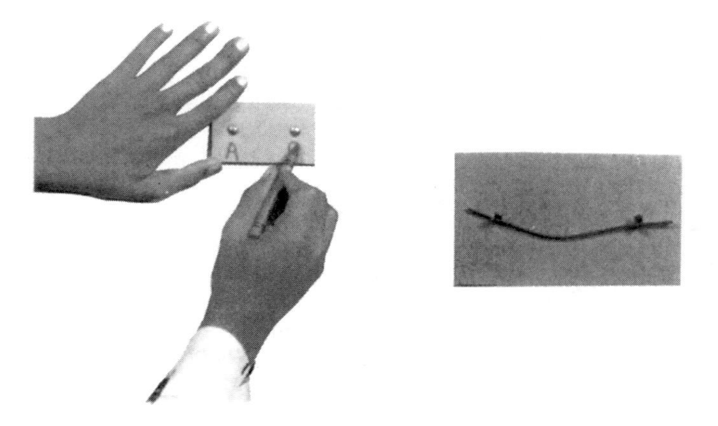

Now, use your circuit tester. Put wire A of the circuit tester and wire B of the circuit tester on the paper fasteners labeled A and B. What happens?

Remove the wire from between the paper fasteners. Then put wires A and B of the circuit tester on the paper fasteners. What happens?

What was the purpose of the wire between paper fasteners A and B? Make a record of your answer.

The picture shows a circuit board. There are six paper fasteners on it. Some of the fasteners are connected with wires. Some are not. The back of the circuit board is covered, and the edges are sealed.

Your teacher will pass out some circuit boards like the one in the picture. *Do not take the back covers off the circuit boards.*

Test the circuit boards to find out which of the paper fasteners are connected with wires. Use your circuit tester. Draw a model of the back of each circuit board you test. Design a way to record your data before drawing the model. You should test at least four different circuit boards.

Design a way to close the circuit tester. Put a device between wire A and wire B that can be used to open and close the circuit. You will be making a *switch*.

If you finish your circuit board before the others, try this experiment: Get two circuit boards. Tie one wire from the top of any brass fastener to any other

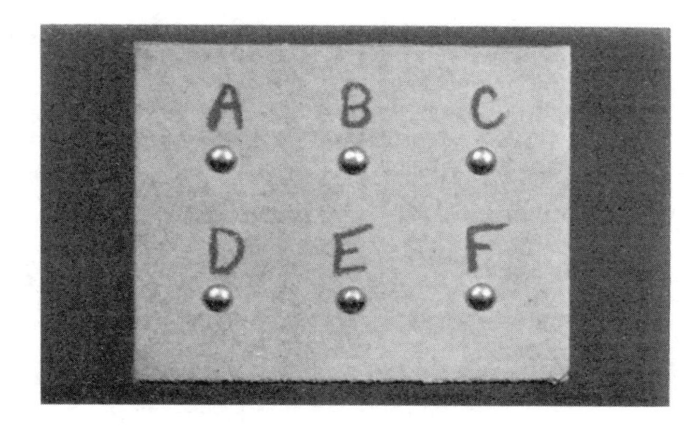

fastener on the other board. Then use your circuit tester. Find out how adding that wire changed the circuits in the board.

Now, add another wire that joins the two boards. What happens to the number of circuits in the two boards after the two wires are added?

Checking Up

1. What are the parts of an electrical circuit?
2. What happens to an electrical circuit when the energy source is removed?
3. What is the difference between a closed circuit and an open circuit?
4. What happens to a closed circuit when you move a switch?
5. Why is the moving of energy from a dry cell to a lightbulb called an electrical current?

Teaching Suggestions The following is a list of materials needed for this learning cycle: brass clips, dry cell, dry cell holder, lightbulbs, lightbulb socket, cardboard, and wires.

The children will be observing the interaction between a dry cell and a lightbulb in order to produce a light. The children will assemble a circuit for the first time, but do not use the word *circuit* at this point. Provide ample time for the children to assemble the circuit. Insist that each child draws a picture of how the wire, dry cell, and lightbulb are connected. To make the lightbulb light, they had to touch the dry cell at its top and bottom. They had to touch the light bulb at its side and at the metal tip on the bottom. Both the dry cell and the lightbulb had to be touched in two places. When this is done, the lightbulb lights.

Before you continue, be sure that each child knows how to make the bulb light. Insist that the children refer to the dry cell as a dry cell and not as a battery. A battery is a collection of dry cells. The children are likely to use the term "battery." Point out that what they actually are using is a dry cell, not a battery.

The electrical current is the interaction of the wire, the dry cell, and the lightbulb. Something happens to the circuit, and that interaction causes electrical current. Do not try to define an electrical current as a flow of electrons or something like water moving along the wires. It is simply the interaction that occurs among the objects in the circuit.

The circuit-board testing will take some time, and you should allow the children to work at their own pace. The students need only to follow the directions in the learning cycle. Allow as much social interaction as possible. Permit the children to share data, exchange ideas, and help each other. This activity is excellent for developing a laboratory spirit—an attitude that assumes that everyone is doing many experiments and is helping everyone else.

References

RENNER, J. W., AND E. A. MAREK. 1988. *The Learning Cycle and Elementary School Science Teaching*. Portsmouth, NH: Heinemann.

APPENDIX A: LEARNING CYCLES FOR PRE-K AND KINDERGARTEN

The pre-K and kindergarten science learning cycles achieve a twofold objective. The program (1) develops children's early understanding of science and (2) introduces process skills that children will be applying to their learning in all the disciplines throughout their years in school.

As an early childhood teacher, you are aware of your students' unique qualities. Your students have learning characteristics that are wholly unlike those of older children. Most certainly in your classroom you will have many preoperational children. In fact, there is a great probability that all the children in your classroom are preoperational. What can preoperational children do?

The preoperational child learns about the environment through *action*. The concept of action means exactly what the word conveys—the child does something physical to the object, event, or situation in the environment. Suppose, for example, that the concept of "rough" was to be taught to kindergarten children. The exploration phase of the learning cycle would consist of letting the children explore all types of rough materials—walls, wood, metal, sandpaper, and other familiar objects. Next, the teacher would intervene and guide the children toward understanding that all the objects they had experienced were alike in one way; all the objects were rough.

The newly constructed concept (the concept of roughness) can be applied to further activities (Applying the Idea). The children can feel other objects that are rough (bark, a stone, a nail file). They can observe the difference between a rough board and rough sandpaper.

The children develop the concept; they acquire knowledge of roughness by acting upon the objects being studied. The learning cycle is appropriate for teaching kindergarten children, but it must be implemented on the action level. Preoperational children cannot yet internalize action and perform mental operations.

Learning Cycle For Pre-K and Kindergarten: Summer and Fall

Gathering Data

Lead the children in a discussion of how their summer days were different from the days they are now experiencing. Now they are in school, and their days are different in many ways. Focus on the things they were doing during the summer that they are not doing now (e.g., going to the beach, playing outdoor games, going on vacations).

Explain that the class will be making a chart. The chart will tell about the things the children were doing before school started. Have available some pictures that can be placed on the chart. The pictures should depict activities such as swimming, picnicking, gardening, sightseeing, vacationing, playing baseball, and celebrating the Fourth of July. Place the pictures on the chart and post the chart on the bulletin board. Ask the children to discuss the pictures and to relate them to their summertime activities. Work up a lively discussion.

Getting the Idea

Ask the children if they know the name of the time of the year they talked about. [The idea: *At one time of the year the days are warm and we do things like picnicking and gardening. This time of year is called summer.*]

Applying the Idea

Write the word *summer* at the top of the chart. You need not assume that the children will learn to read the word *summer*. However, some of them will be developing reading skills and may be able to do so.

Then, apply the idea by having the children look in magazines for additional pictures that remind them of summer. Give them some suggestions. Suggest that they look for pictures of yards, gardens, trees, and summertime events (e.g., baseball games, waterskiing). Bring up the idea that the clothes we wear in summer are different from the clothes we wear during other seasons of the year. Sunsuits, swim trunks, sheer dresses, thin suits—these are the clothes of summer. Turn your chart into a collage of summertime things and events.

Develop your summer chart fully. Then bring out a second piece of chart paper. Explain that the class will now be developing a chart that depicts the events occurring right now. You might first put up a picture of a classroom scene. The children are in school during the fall season. Instruct the children to look in magazines and newspapers for additional pictures of fall scenes.

Create a second chart, called "Fall." Make a collage of the things and events that are typical of the fall season in your community. Do not suggest that the chart be completed immediately. Instruct the children to watch for changes. They can add pictures as the events occur. Among the changes the children might observe are the colorful fall leaves, brown grass, bird migration, and cooler temperatures.

Leave the collages up and continue to discuss and expand them. As the year progresses, you will develop the other two seasons in a similar manner. The children will be comparing one season with the other seasons.

Provide the children with drawings of scenes that are characteristic of different seasons. First, ask the children to color the pictures that make them think of summer. Then, ask the children to look at the drawings of scenes that are characteristic of the other seasons.

As the children study each season, ask them to color the pictures that make them think of that season. In time, they will have made a study of all four seasons.

Learning Cycle For Pre-K and Kindergarten: The Shape of Things

Gathering Data

Give each child a pencil and a piece of paper. Explain that you are going to do something on the chalkboard and that the children are to do the same thing by using their paper and pencil. Place your left hand, with the fingers separated, on the chalkboard and draw around it with a piece of chalk.

Help the children to do the same thing, using paper and pencil. When the outlines are drawn, instruct the children to color them with a crayon. The outlines will show up clearly when colored.

Getting the Idea

Ask the children what they have made on their paper. Ask the children how they could tell their drawing was their hand. [The idea: *We can tell what an object is by its shape.*]

Applying the Idea

Apply this idea by drawing the outlines of other objects on the chalkboard. Put freehand drawings of a cup, a foot, a chair, and an ice cream cone on the chalkboard. Ask the children to identify the objects you have drawn. Emphasize that they are looking at the *shapes* of things.

Show a large paper circle to your class. Say, "Here is another shape." If the children cannot identify the shape, explain to them that it is a *circle*. Ask the children if they can make circle shapes on pieces of paper. Distribute round jar lids and other round objects. Have the children draw around them. Have them color the circles they draw.

Use a similar procedure to introduce the square. Draw a square on the chalkboard by tracing the sides of a square block. Then have the children trace a square block. Discuss sides and corners. Use the words *shape*, *circle*, and *square* as you discuss once again the outlines you have put on the chalkboard.

Have a "shape hunt" in the classroom or on the playground. Ask the children to find objects that have the shape of a circle or a square. If rectangular objects are identified as squares, call attention to the difference in sides. Say that there is a special name for shapes like these—the name *rectangle*.

Show the class a large piece of paper cut into the shape of a triangle. Draw a triangle on the chalkboard. Ask the children to draw the same shape on pieces of paper. Without a triangle to trace, their drawings may be crude, but the children should have little difficulty in conceptualizing the shape. Have your pupils compare the sides of a square with the sides of the figure they have drawn. Make sure they conclude that the new shape has *three* sides, whereas a square has *four* sides. Explain to the class that the shape they have drawn is known as a *triangle*.

For additional application, take the children outside and ask them to arrange themselves to form a circle. Holding hands and standing in a circle is fairly easy. But talk with the children about any "bumps" or "dents" that distort the roundness of their circle. Stress that a circle is round. A true circle has no bumps or dents in its side.

Next, have the children form a square. Have the children discuss how they should make a square. Place a large paper shape on the ground as a model for the arrangement. Point out that all the sides of a square are the same in length. Equal numbers of children on each side make an acceptable formation.

Discuss how the sides join. Use the word *corner* and let the children arrange their lines to form the corners. This is a good time to demonstrate straight lines and crooked lines. Talk about the straight line of the sides of a square. Ask the children if their lines are straight or crooked.

Do not expect anything near perfection in the children's formations. From their positions, the children cannot get a good view of the shape they are making. Suggest that the children squat down. Then permit one or two children at a time to stand up and observe the shape they have made.

Provide the children with drawings of different shapes. Ask the children to color the circles blue. Then ask them to color the squares red and the triangles green.

Strengthen the children's feel for shapes by having them draw shapes in sand. Show how a shape can be drawn in the sand in a sandbox. Ask the children to

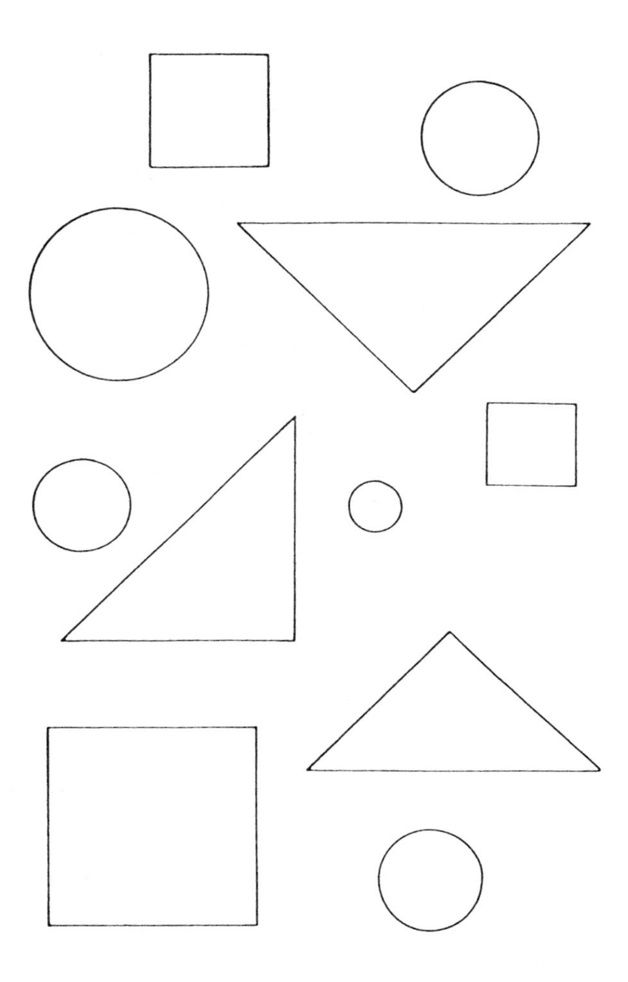

draw shapes with their fingers. Or use round cans and square boxes to make imprints in the sand. Alternate the shapes and discuss how the repeated shapes make a pattern.

Learning Cycle For Pre-K and Kindergarten: Animals at Home

Gathering Data

Begin by talking with children about the houses in which they live. Some people live in houses and others live in apartments. A family may live in a small or large house or apartment. But, no matter how big or small, the place where people live is called a *home*. What about other animals? Where do they go when they go home? What do their homes look like? What are they called?

The children probably know about a bird nest, a fish bowl, a spider web, and an ant hill or colony. They probably know about a stable, a pig pen, and a chicken coop. But do they know about a gopher's burrow or a squirrel's hollow log? Refer to the animals listed below. Ask the children to tell where each animal lives and give the name of its home.

Animal	**Home**
Pig	Pen
Dog	Kennel
Fox	Den/Lair
Cow	Barn
Horse	Stable
Chicken	Coop
Lion	Den
Gopher	Burrow
Ant	Hill/Colony
Bee	Hive
Hornet	Nest
Beaver	Dam
Squirrel	Hole/Hollow Log
Bird	Nest
Spider	Web
Bear	Cave/Den
Goldfish	Bowl
Sheep	Fold
Hermit Crab	Shell

Take the children out on the school grounds and look for animal homes. Take pictures of different animal homes. Bring the pictures back into the classroom. Ask the children to describe the animal homes in the pictures. Show the children pictures taken previously of other animal homes in other locations (e.g., near a lake, in a forest, on a city street).

Getting the Idea

Ask the children to compare the different animal homes they observed. For example, how were the animal homes alike? How were the animal homes different? [The idea: *There are many different kinds of animal homes.*]

Applying the Idea

Along with the idea of this learning cycle, help the children develop the understanding that animals of the same kind live in homes that are alike. For example, all bees live in hives. Ask the children to look at the pictures of animal homes shown.

Then ask the children to look at the pictures of animals shown.

Ask the children to cut out the pictures of animal homes. Then ask them to place the proper home next to the picture of the animal that may live there. Give the children plenty of time to cut out the pictures and paste them alongside the pictures of the animals. Then ask several children to display the work they have done. Evaluate the children's work and make sure that each child has matched the pictures correctly.

Make a class collage from pictures that the children bring to school. Have the children paste their pictures of animals and their homes on a piece of white butcher paper about six or eight feet long. The children can work on the project together, trimming and pasting their pictures.

APPENDIX B: PROTOCOLS FOR FORMAL OPERATIONAL TASKS

The interviewing protocols included here measure the degree of concrete or formal thought a student can use. These protocols have been taken from the work of Jean Piaget and Barbel Inhelder. The adaptations of the tasks to apparatus and indigenous conditions have been made by the authors and other persons associated with the Science Education Center, University of Oklahoma. Any items in the protocols that are at variance with formal thought should be accredited not to Piaget and Inhelder, but to those interpreting how their tasks should be used in interview situations. The following tasks have been presented in this appendix:

1. Conservation of Volume
2. Combinatorial Reasoning
3. Correlational Reasoning
4. Exclusion of Irrelevant Variables
5. Probabilistic Reasoning
6. Proportional Reasoning
7. Separation of Variables

Conservation of Volume

Materials and Procedures

The conservation of volume task was developed by Piaget, Inhelder, and Szeminska (1960). The materials needed for this task include two test tubes and materials that will hold them vertically; two metal cylinders, *exactly* the same size but different in weight, that will fit into the test tubes; strings to lower the metal cylinders into the test tubes; water; and a medicine dropper.

The materials are placed in front of the student. The interviewer should explain the task as follows:

"Here are two test tubes that are exactly the same, and the level of water in them should be the same. Make sure you are satisfied that the water levels are the same in both tubes. If you need to change one of the levels you may use the medicine dropper to make them equal.

"Here are two metal cylinders that are the same size. You will notice one is just as big around and just as tall as the other. [Demonstrate that the cylinders are the same size.] Hold one of them in each hand and tell me how they are obviously different. [The answer should be *weight*.]

"We are going to put one cylinder in each tube. Each cylinder will sink all the way to the bottom of its tube. What will happen to the level of the liquid when the cylinders are submerged in the tubes?" [The answer should be that the water levels will *rise*.]

1. The interviewer then asks: "Will the heavier cylinder raise the water level more? Will the lighter cylinder raise the water level more? Or will both cylinders raise the water levels the same?"
2. The interviewer asks why the student believes as he does, and has the student explain the answer chosen.
3. The interviewer lowers the lighter cylinder into a tube.
4. The interviewer then has the student lower the heavier cylinder into a tube and observe the water level.
5. If the student predicts incorrectly (or correctly, but giving the wrong reason), the interviewer asks what the student thinks caused the levels to come out equally.

Scoring

The scoring procedure for this task is a modified form of the scoring method used by Hensley (1974). In particular, the limits of the total number of combinations necessary to pass the task at each of the four developmental sublevels IIA, IIB, IIIA, and IIIB were adopted from the Hensley scoring method.

IIA: The student makes an incorrect prediction or predicts correctly and gives the incorrect reason; cannot explain the results when she sees the experiment performed.

IIB: The student makes an incorrect prediction or predicts correctly and gives the incorrect reason (as in IIA); however, when the student sees the experiment performed he realizes the correct explanation.

IIIA: The student predicts correctly and gives a correct reason.

IIIB: This task does not require IIIB level of thought.

Combinatorial Reasoning

Materials and Procedures

The colored beads task was designed by Piaget and Inhelder (1951) to test for the presence of combinatorial reasoning. The task was considered to be content-free, since working with materials such as several colored beads does not depend on the students' backgrounds and major areas of specialization. Evaluations of combinatorial reasoning on the colored beads task have also been made by Hensley (1974) and Kishta (1978).

The materials needed for this task include five sets of colored plastic beads: blue (B), green (G), orange (O), yellow (Y), and white (W). The student is presented with the five sets of colored beads and has the opportunity to generate all possible combinations of beads, taking them two at a time, three at a time, four at a time, or all five at a time, as listed below. The interviewer must check to be sure that each set contains beads of the same color, and must be sure that the sets of beads are placed in front of all the students in the same order (B, G, O, Y, W). To generate all possible combinations requires the successful use of combinatorial reasoning. Regarding the function of the involving factors in the combinations, there is no neutral or inhibitor element in the colored beads task. In the colored beads task *all* combinations are considered to be the right answer.

Combinations of Five Colored Beads

Blue (B), Green (G), Orange (O), Yellow (Y), and White (W)					
BG	GY	BGO	BYW	BGOY	BGOYW
BO	GW	BGY	GOY	BGOW	
BY	OY	BGW	GYW	BGYW	
BW	OW	BOY	GOW	BOYW	
GO	YW	BOW	OYW	GOYW	

The protocol for this task is a modified form of the protocol developed by Hensley (1974). The rack of five sets of beads is placed in front of the student and the interviewer says, "Here are five plastic containers, each containing beads of different colors. Your task is to make groups of beads. A group has two or more beads of different colors. Also, a group will not have more than one bead of the same color. The order in which you place the beads makes no difference. Blue and orange, orange and blue are the same. Using as many beads as you wish, I would like you to make as many different groups of beads as you can. Do you have any questions?" After the student responds the interviewer says, "You may begin." During the experiment the interviewer uses questions such as the following:

1. "Have you made all possible groups of beads?"

2. If the answer is no, the interviewer says, "Please keep trying to make as many different groups as you can."

3. If the student stops again, the interviewer goes back to the first question.

4. If the answer is yes, the interviewer says, "all right," and the experiment terminates at this point.

Scoring

IIA: The student is able to make most of the bead pairs (combinations of twos). She might combine the beads three at a time, four at a time, or all five at a time. Nevertheless, the IIA student does not complete more than three combinations of higher orders.

IIB: The student is able to generate the ten bead pairs or all of the beads together. He completes some higher combinations without any systematic approach, but no more than eighteen total combinations.

IIIA: The new innovation that appears at substage IIIA is the introduction of systematic method in the use of $n \times n$ combinations. The IIIA student completes between nineteen and twenty-two combinations in a systematic way. The minimum number of combinations to pass the task at IIIA level was also reported to be nineteen in Hensley (1974) and Kishta (1978).

IIIB: The student has the attributes of substage IIIA, but combinations are organized in a more systematic fashion from the start and with greater speed. At substage IIIB, the student generates more than twenty-two combinations.

Correlational Reasoning

Materials and Procedures

We extend our thanks to Dr. Anton E. Lawson, Arizona State University, for his assistance in preparing this protocol. The materials needed for this task include sets of cards with fat or thin rats with green or red tails. The numbers shown represent the number of cards required for each trial.

Trial 1: 7 fat rats with red tails
2 fat rats with green tails
3 thin rats with red tails
8 thin rats with green tails

Trial 2: 2 fat rats with red tails
8 fat rats with green tails
8 thin rats with red tails
3 thin rats with green tails

Trial 3: 6 fat rats with red tails
9 fat rats with green tails
4 thin rats with red tails
6 thin rats with green tails

The interviewer begins with the following introduction: "A scientist recently conducted an investigation of the rats that inhabit a series of small islands in the South Pacific. On the first island she visited, she found that all the rats were either quite fat or quite thin. [Show examples to the student]. Also, all the rats had either green tails or red tails. This made the scientist wonder if there might be a connection or a relation between the size of the rats and the color of their tails. So she decided to capture some of the rats and observe them to see if there was some connection."

The interviewer hands the first of three sets of data cards to the student, then asks the student to examine the cards and check to see if there is, in fact, a relationship. Students initially unable to proceed may be given assistance in classifying the cards into four categories. The interviewer allows a few moments for the student to explore and categorize the cards. The interviewer may ask, "Are there more fat rats than thin rats?" "Are there more rats with green tails than with red tails?" "Do you find any connection between the thin rats and the tail color?" The student then is asked to explain any relation found. If the student shows some success on the first set of data, two additional trials may be presented.

Scoring

IIA: Multiple classification of incidents is implied in the student's response. The student describes various events qualitatively: "I don't know, because there are some rats of each kind"; "I think there is no connection. There are seven rats with green tails and three with red tails and there are two thin rats with green tails and eight with red tails."

IIB: Multiple classification of incidents is implied, and the incidents are compared in the student's response. The number of cases is compared: "The island made them have different colors and shapes, because there are more rats with green tails than red, and more fat rats than thin."

IIIA: Beginning of correlational reasoning using probabilities or proportionalities appears in the student's response. The student compares the incidences in two pairs, "There are more red fat than green fat, and more green thin than red thin rats."

IIIB: A satisfactory solution of the problem with a clear identification of the mathematical relationships being used for the comparisons is apparent in the student's response. The student identifies and/or compares two ratios: "More small rats have green tails by eight to three, and more large rats have red tails by seven to two"; "With red tails the rats are more likely to be small—about 80 percent—and with green tails the rats are most likely to be big—70 percent"; the student compares the number of confirming cases or disconfirming cases.

Exclusion of Irrelevant Variables

Materials and Procedures

The pendulum experiment was developed by Inhelder and Piaget (1958) to measure students' ability to exclude irrelevant variables. The materials needed for this task include a ring stand with ring; three strings of different lengths, and three weights (50, 100, and 150 grams).

The interviewer places the materials in front of the student and says, "Here are the materials to make a pendulum. You may hang a string on the stand, place a weight on the string, and swing the pendulum. [Show the student.] I would like you to work with the materials to determine what controls how fast the pendulum swings." Ask the student to identify the variables. If students cannot name all of the possible variables (i.e., push, drop, weight, length of string), point them out for them.

During the experiment the interviewer uses questions such as the following:

1. Have you tried all of the possible ways to change the speed of the pendulum swing?
2. What things have you changed to see if the pendulum swing speed changes?
3. What factors control the rate at which the pendulum completes its swing?
4. Why did you choose the (one of the four factors) to test?

Scoring

IIA: Variations in the motion of the pendulum attributed to the push it is given when starting. The length of the string may be mentioned as somewhat affecting the rate, but is not singled out. Several variables are varied simultaneously.

IIB: Variations in the motion of the pendulum are explained as an inverse relationship between the length of the string and the motion of the pendulum. Variables are not isolated. "The longer the string, the slower its motion." "The shorter the string, the faster its motion."

IIIA: The student separates the variables, such as the length of string, the size of the weight, the height from which the weight is released, and the force of the push, but the student has difficulty in controlling each variable while experimenting. A hypothesis may be formed that variation is caused by the length of the string, yet the student will qualify its role.

IIIB: The student separates the variables and formulates a hypothesis that leads to the exclusion of all factors except the length of the string.

Probabilistic Reasoning

Materials and Procedures

The materials needed for this task include squares (four yellow, five blue, three red), diamonds (two yellow, three blue, seven red), and a box.

The interviewer provides a box to which blue, red, or yellow squares and diamonds are added. The interviewer begins by explaining to the student that different numbers of squares and/or diamonds will be added to the box. The student will be asked to solve several problems. The interviewer continues:

1. Put *two yellow squares* and *two blue squares* in the box. Shake the items and ask, "What are the chances of my drawing a *blue square* on the first draw?"
2. Put *one each of red, blue, and yellow squares* and *two each of red, blue, and yellow diamonds* into the box. Shake the items and ask, "What are the chances of my drawing a *square of any color* on the first draw?"
3. Put *five blue squares, four yellow squares*, and *three red squares* into the box with *three blue diamonds, two yellow diamonds*, and *four red diamonds*. Shake the items and ask, "What are the chances of my drawing a *red piece (either a square or a diamond)* on the first draw?"
4. Using the same items as in number three, ask, "What are the chances of my drawing a *blue diamond* on the first draw?"
5. Using the same items as in number three, add *three more red diamonds* and ask, "What are the chances of my drawing out a *yellow square* on the first draw?"

Scoring

IIA: The student is unable to answer the first question.

IIB: The student is able only to give the 2:1 proportion response to the first question.

IIIA: The student identifies a 1:3 proportion.

IIIA: The student states that a 1:7 proportion is present.

IIIB: The student identifies the 1:6 proportion when all other proportions are identified correctly.

Proportional Reasoning

Materials and Procedures

The materials needed for this task, called Equilibrium in the Balance, are listed by Inhelder and Piaget (1958).

The apparatus is placed in front of the student. The interviewer says, "Here is a bar that is balanced at the center. Notice that there are seventeen evenly spaced hooks on each side of the balance point on which to hang weights. We are going to do some balance tasks. I'll hang a weight on one side and ask you to balance the bar by hanging a weight or weights on the other side. You can hang weights on different hooks or on the same hook if you wish. The weights can be hooked together to suspend them from the same hook."

The interviewer proceeds with the following steps:

1. "First, I'm going to hang a 100 gram weight on the sixth hook." [Count from the center.] "I want you to place a 100 gram weight on the other side to make the bar balance." The interviewer holds the bar level while the student is hanging the weight, and before releasing it asks this question: "Why did you hang the weight on the _______ hook?"

2. "I'm going to leave my 100 gram weight on the sixth hook and give you two 50 gram weights. Where will you hang your 50 gram weights to balance my 100 gram weight?" The interviewer holds the bar level while the student decides where to place the two 50 gram weights and places them. Before releasing the bar, the interviewer asks, "Why did you hang the 50 gram weights on the _______ hook (or hooks)?"

3. "Now, I'm going to hang a 100 gram weight on the sixth hook." [Count from the center.] "I want you to take a 50 gram weight and hang it on the other side to balance the bar." The interviewer holds the bar level while the student is placing the weight. While holding the bar level, the interviewer asks: "Why did you hang the weight on the _______ hook?"

4. "Next, I'm going to hang a 120 gram weight on the third hook." [Count from the center.] "Hang a 40 gram weight on the other side to make the bar balance." The interviewer holds the bar level while the student is placing the weight. Still holding the bar level, the interviewer asks, "Why did you hang the weight on the _______ hook?"

5. "Now I'm going to hang a 70 gram weight on the tenth hook." [Count from the center.] "Hang a 100 gram weight on the other side to make the bar balance." The interviewer holds the bar level while the student is placing the weight. Still holding the bar level, the interviewer asks, "Why did you hang the weight on the _______ hook?" If the interviewer is unsure of his judgment regarding the interviewee's success after using the 70/100 weight combination, a good way to test it is to place a 60 gram weight on the sixth hook and give the student 40 grams. The procedure followed is the same as that given above.

One amendment can be made to this task if desired. The magnitude of the weights and distances can be changed, provided that the proportions of 1:1, 1:2, 1:3, and a more complex one (e.g., 3:2, 7:10) are maintained.

Scoring

IIA: The student is not successful with anything beyond step two.

IIB: The student is successful with the two-to-one proportions of step three. The explanation must include the use of the proportion concept.

IIIA: The student is successful in balancing the bar using the weights and distances outlined in step four. The explanation must include the proportion concept.

IIIB: The student is successful in balancing the bar using the weights and distances outlined in step five. The explanation must include the proportion concept. A student who solves the problem using a rule such as *weight times distance on one side equals weight times distance on the other side* is using an algorithm, without necessarily using a proportion. If this is evident, the student is told, "Give me another solution using weight and distance in some other way." If the student cannot satisfactorily explain this using proportions, a lower score is given, depending on the last level for which a satisfactory explanation was given.

Separation of Variables

Materials and Procedures

The materials needed for this task, called the bending rods task, are listed by Inhelder and Piaget (1958).

The apparatus is placed in front of the student and the interviewer uses an explanation such as the following:

"I have an apparatus and I want to show you how it works. I can pull the rods back and forth and make them as long or as short as I want. The effective length is from here out." [Demonstrate.] "The screws must be loosened in order to move the rods." The interviewer allows the student a few moments to explore the apparatus, and makes sure the student understands that adjusting the lengths of the rods is permitted.

The interviewer says, "Look at the rods and tell me in as many ways as you can how the rods are different." The interviewer leads the student to state the three ways the rods are different and explains that these differences are called variables. If the student does not find all the variables, the interviewer explains what they are. At this point the weights are introduced, and the student is shown how the weights will bend the rods. The interviewer now restates the four variables.

The interviewer next says, "Do some experiments to show me the effect of each one of the four variables on how much the rods bend." The interviewer is inviting the student to demonstrate at least one experiment to show the influence of one of the variables on the bending of the rods. That is a category IIIA characteristic.

If the student does not reach the IIIA level, the interviewer should provide the opportunity to reach the IIB level. A good instruction to lead back to IIB is this: "Take one thick rod and one thin rod and make them bend the same amount, using two equal or identical weights." The student in category IIB solves this problem by logical multiplication and explains why. The intuitive feeling is present that long-thick balances short-thin. The IIA student does not demonstrate logical multiplication.

After one experiment controlling variables has been done and after the student is established at the level IIIA, a good question to use in leading the student is this: "What else can you do to test the other variables?" The interviewer may precede this question by asking, "There are three more variables. Do you remember what they are?"

A good question to lead the student during the task, but not before the student has had the opportunity to set up an experiment, is this: "What can you do to prove that the material (or length, or diameter) of the rod is important in determining how much it will bend?" The following questions should be asked throughout the interview, after each experiment the student attempts: "What are you showing with that experiment?" "What variable is your experiment dealing with?" "How does your experiment show what variable you are testing?" A good question to conclude the interview for this task is this: "Is there anything else you want to do with this apparatus?"

Scoring

IIA: The student cannot explain logical multiplication.

IIB: The student can explain logical multiplication (intuitive feeling that long and thick balances short and thin).

IIIA: The student does at least one experiment that proves the effect of at least one variable.

IIIB: The student solves the entire problem.

References

HENSLEY, J. H. 1974. "An Investigation of Proportional Thinking in Children from Grades Six Through Twelve." Unpublished doctoral dissertation, University of Iowa.

INHELDER, B., AND J. PIAGET. 1958. *The Growth of Logical Thinking*. New York: Basic Books.

KISHTA, A. E. 1978. "Proportional and Combinatorial Reasoning in Two Cultures." *Journal of Research in Science Teaching* 15:11–24.

PIAGET, J., AND B. INHELDER. 1951. *The Origin of the Idea of Chance in Children*. New York: Norton.

PIAGET, J., B. INHELDER, AND A. SZEMINSKA. 1960. *The Child's Concept of Geometry*. New York: Harper and Row.

INDEX

abstract
  concepts, 11, 12, 13
  thought, 54
abstracting concepts, 10–11
accommodation
  defined, 60–61
  fostering, 108–109
actions and operations in learning, 36–37
actual developmental level, 94–95
adaptation, 61–62
Alling, Mary, 17
alternative assessments
  concept maps, 147–148
  creative writing, 145–146
  illustrations, physical models, and analogies, 143–144
  journals and learning logs, 148
  aboratory practicals, 146–147
  mental models, 144–145
  observations, teacher, 149
  oral tests and interviews, 144
  overview, 150
  portfolios, 149
  projects and presentations, 148–149
  science process assessments, 142–143
analogies, assessment using, 143–144
analyzing, 19, 28–29
animals at home, sample pre-K and kindergarten learning cycle for, 227–229
aquarium building, sample learning cycle for, 164–167
area, conservation of, 47–48
art
  integrating, in science learning cycle, 122
  learning cycles for, developing, 123, 125–126
assessment
  alternative assessments. *See* alternative assessments
  conventional tests, 149–150
  overview, 141–142
assimilation
  conditions necessary for, 107
  defined, 59
  in learning cycle, 69–70
  role of, in mental functioning, 59–63
assumptions, reasoning from, 55
Ausubel, David, 97–99

Behind the Methods Class Door: Educating Elementary and Middle School Science Teachers, 117
Biological Science Curriculum Study (BSCS), 14
biological sciences
  concepts from, 153–156
  learning cycles, sample, 164–182, 227–229
Bohr, Niels, 4, 9

Camp, Katherine, 21–22
CDs, 135
centering, 42, 44–45
Central Purpose of American Education, The, 18
Children Exploring Their World: Theme Teaching in Elementary School, 123
circulation, sample learning cycle on human, 178–182
classifying, 19, 27–28, 52
classroom safety, 137–140
Clay Boats: Experiments with Sinking, Floating and Simple Volume Relationships, 117
clues, teacher use of, 95–97
cognitive levels of questions, 129–130
collaborative groups, 96, 97
combinatorial
  reasoning, 232–233
  system, 55
community, concept of, 82, 83

compact discs, 135
comparing, 19, 25–26
computers, 135–136
concept-application phase
  demonstrations used during, 133
  developing learning cycles and the, 109–111
  questions to ask during, 130
  role of, in learning cycle, 8–9, 70, 74–75, 100
concept maps, assessment using, 147–148
concepts
  abstracting concepts, 10–11
  from biological sciences, 153–156
  concept-application phase and, 109–111
  concrete, 11, 12, 53, 105–106
  content, 73
  defined, 10–13
  from earth sciences, 156–160
  formal, 55, 106
  meaningful learning theory and, 97–100
  misplacement in the school, results of, 13
  from physical sciences, 160–164
  selecting, to be taught, 105–106
  "teaching concepts," 11
  term introduction phase and, 108–109
conceptualization, 11
concrete concepts, 11, 12, 53, 105–106
concrete operational stage, 52–54, 60, 63
  concept selection for, 105–106
concrete reasoning, 53, 54–55, 230–239
conservation reasoning, 42, 45–51, 52, 230–232
constructivism, 15, 35
content
  concepts, 73
  culturally imperative, 110
  factor in learning, 63–64, 66
  rational power development and, 20–21
controlled questions, 129
conventional tests, use of, 149–150
cooperative learning, 96, 97, 135, 136
correlational reasoning, 233–235
correspondence, 52
creative writing, assessment using, 145–146
culturally imperative content, 110
culture, 94

data
  collection, computers used for, 135
  interpretation activities, 81–85
  processing data from the environment, theory on, 59–63
decomposition, sample learning cycle for, 176–178
deducing, 19, 27
deferred imitation function, 40
demonstrations, use of, 132–134
developing learning cycles. *See also* learning cycles
  concept application, 109
  concept selection, 105–106
  exploration, 106–108
  guidelines for, 111–114
  integrating science with other subjects, 116-123
  non-learning cycle materials, using, 114–115
  for other subjects, 123–127
  students' guides, suggestions for preparing, 112, 114
  teachers' guides, suggestions for writing, 112, 113
  term introduction, 108–109
  what is to be taught, finding, 105–106
  what to do, deciding, 106–111
developmental learning model, 58
development and learning, relationship between, 71–73
direct experience, 12, 13
discrepant event, 114–115, 132
disequilibrium, 60–61
drawing function, 41

earth sciences
  concepts from, 156–160
  learning cycles, sample, 183–201, 224–226
Educational Policies Commission (EPC), 18, 19, 20, 21
egocentrism, 41, 42
Einstein, Albert, 3, 4, 9
electricity, sample learning cycle for, 216–222
energy, sample learning sample for, 214–216
English as a second language (ESL) students, 136
environment, processing data from, 59–63
equilibrium, 60–61
errors, dealing with student, 84–85
essential experiences of science
  experimenting, 85–86
  fitting, into school, 92
  interpreting, 81–85
  and learning cycle, relationship of, 76–78
  measuring, 79–81
  model building, 87–89
  observing, 78–79
  predicting, 89–91
evaluating, 19, 31–32
evaluation. *See* alternative assessments; assessment
exclusion of irrelevant variables, 235–237

experience, influence of, 56
experimentation, 85–86
experiments, in traditional teaching practice, 4
explanation, 87
exploration phase
  demonstrations used during, 132–133
  developing learning cycles and the, 106–108
  questions to ask during, 130
  role of, in learning cycle, 5, 7, 9, 70, 74
exposition phase, 3
expository teaching procedures, overuse of, 20

facts, 10
fall and summer, sample pre-K and kindergarten learning cycle for, 224–225
food chain, sample learning cycle for, 168–171
formal
  concepts, 55, 106
  reasoning, 53, 54–55, 230–239
formal operational stage, 54–55
fossils, sample learning cycle for, 189–193
free assimilation, 107
fronts, sample learning cycle for weather, 198–201
Full Option Science System (FOSS), 14–15
function, intellectual, 62–64, 70

generalization, 10–11
generalizing, 19, 26–27
geography in science learning cycle, integrating, 121
getting food, sample learning cycle for, 168–171
goals
  of schools, 18–20
  of science education, 20–22
goals of science education, 21–22
grouping
  assessment using, 148–149
  value of, 96, 97
growing plants, sample learning cycle for, 167–168
guided assimilation, 107

heart, sample learning cycle on working of the, 178–182
histogram, 172, 173, 174
history in science learning cycle, integrating, 121
Huxley, Thomas, 21
hypothesizing, 27, 73
hypothetico-deductive thought, 54

illustrations, assessment using, 143
imagining, 19, 29–30
indirect evidence, 12
individual projects, assessment using, 148–149
inductive thinking, 27
inferring, 19, 26
inform-verify-practice (IVP) procedure, 3–5, 84
Inhelder, Barbel, 230–233, 235–239
injuries, 140
inquiry
  defined, 19–20, 21
  using, to develop rational powers, 20–21
intellectual development
  concrete operational stage of, 52–54, 60, 63
  educational practices and, 55–58
  environment, processing information from the, 59–63
  formal operational stage of, 54–55
  and intelligence, relationship between, 65
  mental structures and, 59–63
  movement through levels, 55–58
  Piaget's theory, evolution of, 37–38
  preoperational stage of, 39–52, 59–60, 62, 71–73
  sensorimotor stage of, 38–39
intellectual function, 62–64, 70
intelligence
  and intellectual development, relationship between, 65–66
  invariants of, 62
  Piaget's model of, 64
interiorized action, 37
interpreting activities, 81–85
interviews, assessment using, 144
invariants of intelligence, 62
irrelevant variables, exclusion of, 235–237
irreversibility, 41, 42–44
Island of the Blue Dolphins, 122
Island of the Skog, The, 122

journals, assessment using, 148

Karplus, Robert, 14
Kettleship Pirates, The, 122
Kleine, Sharlene, 137
knowledge
  actions and operations in development of, 36–37
  construction, 11, 35, 58
  and memorization, 35
Kon-Tiki, 122

laboratory
  journals, assessment using, 148
  practicals, assessment using, 146–147

safety and the, 138, 139–140
"lack-of-coverage" complaint, 76
language arts
developing learning centers for, 123, 126
integrating, in science learning cycle, 121
laser discs, 135
Lawson, Anton E., 233
learning
content factor and, 63–64, 66
defined, 57–58
and development, relationship between, 71–73
Piaget's model of, 59–64
products of, 73
learning cycle
and the ability to think, 73–74
assessment. (*see* alternative assements; assessment)
demonstrations, use of, 132–134
and essential experiences of science, relationship of, 76–78
historical overview, 14–15
"lack-of-coverage" complaint and the, 76
phases, 9, 14, 70
philosophical basis, 15
questioning strategies, use of, 128–131
safety and the, 137–140
and science, 9, 74–75
technology, use of, 134–136
Learning Cycle and Elementary School Science Teaching, The, 152
learning cycles
for biological sciences, sample, 164–182
developing learning cycles. (*see* developing learning cycles)
for earth sciences, sample, 183–201
integrating, with other subjects, 115–123
for kindergarten, 223–229
for physical sciences, sample, 201–222
for pre-K, 223–229
learning logs, assessment using, 148
lecture, overuse of, 20
length, conservation of, 48–49
liquid amount, conservation of, 47, 51
logical-mathematical experience, 56
logical thinking
defined, 19
as educational goal, 18–20
fostering, in science education. (*see* rational powers)
lungs, sample learning cycle for, 171–175

magnetism, sample learning cycle for, 206–207
maps
assessment using concept, 147–148
learning cycle for reading, sample, 193–198
Maps and Globes, 122
mass questions, 129
Materials Safety Data Sheets (MSDS), 138, 140
mathematics
integrating, in science learning cycle, 120–121
learning cycles for, developing, 123–125
matter, sample learning cycle for states of, 208–210
maturation, 56
McGuffin, Tina, 123
meaningful learning theory, 97–100
measurement
learning cycle for, 79–81, 106–108, 109–111
science process questions for, 142–143
value of, activities, 79
memorization
and knowledge, 35, 57–58
mental ability and, 41
overuse of, 20
mental
models, assessment using, 144–145
operations, 36–37
structures, 59–63
mental image function, 41
Merrill, Winifred Edgerton, 5, 9
Mitchell, Maria, 4
mixtures, sample learning cycle for, 210–214
Moby Dick, 122
modeling, 95–97
models
mental, assessment using, 144–145
model building activities, 87–89
physical, assessment using, 143
value of, 12–13
music, developing learning cycles for, 123, 125–126
Music and Moments with the Masters, 126

natural science, 3
nature of science, 3
nonverbal reinforcement, 131
number, conservation of, 46–47, 51

object teaching, 21
observations, assessment using teacher, 149
observing activities, 78–79
on-line services, 136
operations and actions in learning, 36–37
oral
student instructions, 112, 114
tests, 144
organization, 61, 75

peer-assisted learning, 96, 97, 135, 136

physical experience, 56
physically impaired students, 136
physical models, assessment using, 143
physical sciences
  concepts from, 160–164
  integrating a learning cycle in, 117–123
  laboratory practical examination, sample, 146
  learning cycles for, sample, 201–222, 225–227
physiology, integrating a learning cycle in, 116–117
Piaget, Jean
  actions and operations in learning, theory on, 36–37
  data from environment, theory on processing, 59–63
  developmental theory, contributions to, 35
  formal reasoning, measuring, 230–233, 235–239
  intellectual development theory. (*see* intellectual development)
  intelligence model, 64
  learning and development, explanation of relationship between, 71
  learning model, 59–64
plants, sample learning cycle for growing, 167–168
Poincairé, Henri, 4–5
portfolios, assessment using, 149
potential developmental level, 95
practice phase, 4
prediction, 6–7, 89–91
preoperational stage, 39–52, 59–60, 62, 71–73
  teaching methodology and, 105, 106, 109
principle, 11
probabilistic reasoning, 236–237
products of learning, 73
Project 2061, 17, 18
projects, assessment using, 148–149
properties of objects, sample learning cycle for, 202–206
proportional reasoning, 237–238
propositional reasoning, 54
Psychology of Intelligence, 40
purpose of schools, 18–20

questioning
  in social constructivist theory, 95–97
  strategies, use of, 128–131
Quine, David, 125

rational powers
  ability to think and use, 73–74
  analyzing, 28–29
  classifying, 27–28
  comparing, 25–26
  deducing, 27
  defined, 19
  development of, 19, 22–32
  evaluating, 31–32
  fostering, in science education, 20–21
  generalizing, 26–27
  imagining, 29–30
  inferring, 26
  recalling, 22–25
  synthesizing, 30–31
reading in science learning cycle, integrating, 122
reading maps, sample learning cycle for, 193–198
reality, 55
reasoning. *See also* thinking ability
  combinatorial, 232–233
  concrete, 53, 54–55, 230–240
  conservation, 42, 45–51, 52
  correlational reasoning, 233–235
  intellectual development. (*see* intellectual development, levels of)
  measuring, 230–239
  probabilistic, 236–237
  proportional reasoning, 237–238
  rational powers, fostering. (see rational powers)
  transductive, 42, 45, 71–72
recalling, 19, 22–25
reinforcement, 130–131
relevant prior knowledge, 98
Renner, John W., 14
reversibility, 43
reversible action, 37
Robinson Crusoe, 122
Robinson Crusoe Story, The, 122
rocks, sample learning cycle on soil and, 183–185
Roller, Duane, 5
rote learning, 97, 98, 99

safety, 137–140
"Save the Whales," 122
scaffolding, 95
schemes, 59, 72–73
schools, purpose of, 18–20
science
  defined, 3–5
  emphasis on process of, 4–9
  and learning cycle, 9, 74–75
Science Curriculum Improvement Study (SCIS), 14
science education
  assessment. (*see* alternative assessments; assessment)

essential experiences of science. (*see* essential experiences of science)
goals of. (*see* goals of science education)
intellectual development and, 55–58
learning cycles for. (*see* developing learning cycles; learning cycles)
rational powers, fostering. (*see* rational powers)
teaching model for, 69–70
theory base for, 69–75
traditional, 3–5
Science for All Americans, 17
science of nature, 3
science process assessments, 142–143
scientific terminology, 109
self-regulation, 61
semiotic functions, 40–41
sensorimotor stage, 38–39
separation of variables, 238–239
seriation, 52
shapes, sample pre-K and kindergarten learning cycle for, 225–227
signifier, 40
Simpson, William, 116
social constructivist theory, 93–97
social transmission, 57
software, learning cycle and, 135–136
soil, sample learning cycle on rocks and, 183–185
solid amount, conservation of, 47, 51
solitary questions, 129
special needs students, 135–136
spontaneous questions, 129
standardized tests, use of, 150
standard units of measure
learning cycle for, 106–108, 109–111
science process questions for, 142–143
states of matter, sample learning cycle for, 208–210
students' guides, suggestions for preparing, 112, 114
subcultures, 94
subsumers, 98
summer and fall, sample pre-K and kindergarten learning cycle for, 224–225
Swiss Family Robinson, The, 122
symbolic functions, 40–41
symbolic play function, 40–41
synthesizing, 19, 30–31

teacher-led discussion, term introduction with, 7
teacher observations, assessment using, 149
teachers' guides, suggestions for writing, 112, 113
teaching model, 69–70
teaching science. *See* science education
technical skill, using demonstrations to teach a, 133–134
technology, use of, 134–136
temperature, concept of, 11–12
terminology, use of, 9, 109
term introduction phase
developing learning cycles and the, 108–109
questions to ask during, 130
role of, in learning cycle, 7–8, 9, 70, 74
tests, use of conventional, 149–150
textbook, overuse of, 20
thinking ability. *See also* reasoning
abstract, 54
defined, 19
as educational goal, 18–20
learning cycle and, 73–74
rational powers, fostering. (*see* rational powers)
traits in a transformation, 44
transductive reasoning, 42, 45, 71–72
transmission, 57
Treasure Island, 122

University of Chicago Laboratory School, 21

variables, separation of, 238–239
VCRs, 134–135
verbal evocation function, 41
verbal reinforcement, 131
verification phase, 4
video recorders, 133–134
visually impaired students, 136
volcanoes, sample learning cycle on, 187–189
volume, conservation of, 230–232
voluntary questions, 129
Voyage of the Dawn Treader, The, 122
Vygotsky, Lev, 94–96

Walmsley, Sean, 123
weather fronts, sample learning cycle for, 198–201
weight, conservation of, 49–50
written student instructions, 112, 114

year, sample learning cycle on the, 185–187

zone of proximal development, 95